Social Robots from a Human Perspective

Jane Vincent · Sakari Taipale · Bartolomeo Sapio
Giuseppe Lugano · Leopoldina Fortunati
Editors

Social Robots from a Human Perspective

Springer

Editors
Jane Vincent
School of Arts
University of Surrey
Guildford, Surrey
UK

Giuseppe Lugano
COST
Brussels
Belgium

Sakari Taipale
University of Jyväskylä
Jyväskylä
Finland

Leopoldina Fortunati
University of Udine
Udine
Italy

Bartolomeo Sapio
Fondazione Ugo Bordoni
Rome
Italy

ISBN 978-3-319-15671-2 ISBN 978-3-319-15672-9 (eBook)
DOI 10.1007/978-3-319-15672-9

Library of Congress Control Number: 2015935413

Springer Cham Heidelberg New York Dordrecht London

Printed on acid-free paper

Springer International Publishing AG Switzerland is part of Springer Science+Business Media (www.springer.com)

Acknowledgments

Although the robotification of societies is still at an early phase in Europe and worldwide, robots do not leave people cold. They provoke strong thoughts, both for and against. This volume addresses robotics from a human perspective, highlighting the social, cultural, and political aspects of robotics. In so doing, the book attempts to present social robots in proportion by illustrating people's own attitudes and perceptions, as well as ways of how social robots gain a foothold in societies through the increased automatization and digitalization which seem to be leading to the introduction of more and more self-sufficient robots. One message of the book is that social robots do not turn up out of the blue, but many of their social functions are familiar to all of us from the world of digital media. For example, the user interfaces of social robots are designed to be intuitive and user-friendly by utilizing existing mundane technologies, such as tablets, mobile phones, touch screens, and web cameras, as well as all kinds of sensors already embedded in mobile communication devices. This edited volume is born out of the international workshop "The Future Concept and Reality of Social Robotics: Challenges, Perception and Applications. Role of Social Robotics in Current and Future Society" organized by COST (European Cooperation in Science and Technology) in Brussels during 10–13 June 2103. The original idea behind this book was to bring together the most robust empirical research work and the most illustrative of non-empirical social analyses of social robots presented in the workshop. The ten papers included in this volume were drawn from over 76 papers presented in Brussels. The outcome of this effort rests with the reader now.

Finally, we would like to thank COST for making the workshop possible. Any opinions, findings, conclusions, or recommendations expressed in the book are those of the respective author(s), and do not necessarily reflect the views of their background organization or COST. We are also grateful to Springer, especially to Senior Editor Anthony Doyle and Ms. Amudha Vijayarangan, for their commitment and support to this project. Also, all our peer reviewers who assisted in assuring the quality of the chapters, and Ms. Emilia Leinonen, who assisted in the final preparation of the book manuscript, deserve thanks and appreciation.

Jane Vincent
Sakari Taipale
Bartolomeo Sapio
Giuseppe Lugano
Leopoldina Fortunati

Contents

Editors and Contributors

About the Editors

Jane Vincent Ph.D., is Senior Research Fellow at the LSE Media and Communications Department and Visiting Fellow with the Digital World Research Centre at the University of Surrey. She researches the social practices of mobile communications and ICT users. Her studies for industry and international academic organizations on the social shaping of technology, children's and older peoples' use of mobile phones are published widely and Jane's work on emotions and mobile phones is published in English, German, and Russian.

Sakari Taipale Ph.D., is an Academy of Finland Research Fellow at the University of Jyväskylä, Finland. He is also an Adjunct Professor at the University of Eastern Finland. Presently, he is running a five-year research project on intergenerational relations in the digital societies. Taipale has published on the social aspects of new media technologies and mobilities in many high-ranked academic journals, such as British Journal of Sociology, New Media and Society, Information, Communication and Society, Social Science Research, Telecommunications Policy, European Journal of Communication, and Mobilities. He has also edited a book with Prof. Juho Saari and Dr. Sakari Kainulainen. Regarding robot research, Taipale has studied the Eurobarometer 382 "Public Attitudes towards Robots" data collected in 2012 and lectured on social robotics at international academic events.

Bartolomeo Sapio Ph.D., holds a Doctor Laurea Degree in Electronic Engineering summa cum laude at the University of Rome "La Sapienza". He is a Senior Researcher and Project Manager with Fondazione Ugo Bordon participating in several international projects. He has carried out methodological research in the field of scenario analysis, developing the original methods WISE, SEARCH and GIMMICKS, and applying them to multimedia, broadband networks, the Internet, mobile services, and the convergence between fixed and mobile networks. He designed and implemented SIMULAB (Scenario-engineering Interactive Multimedia LABoratory), an advanced work environment to carry out research activities in the field of Scenario Engineering. He was Chairman of COST (COoperation in the field of Scientific and Technical research) Action 298 "Participation in the Broadband Society", and a member in the Management Committees of COST Action IS0605 "A Telecommunications Economics COST Network—Econ@Tel" and COST Action A22 "Foresight methodologies—Exploring new ways to explore the future," where he chaired the Working Group "Integrating Narratives and Numbers". He is currently the co-ordinator of an EU-funded project carried out within the Seventh Framework Programme: CUMULUS (Certification infrastrUcture for MUlti-Layer cloUd Services).

Giuseppe Lugano Ph.D., works in Brussels as Science Officer for COST, the longest-running European framework for cooperation in science and technology. He holds an M.Sc. degree in Computer Science and a Ph.D. in Cognitive Science. His publications and research interests deal with the conceptual design of technologies and services for community and sustainable living.

Leopoldina Fortunati Ph.D., is Professor of Sociology of Communication at the Faculty of Education of the University of Udine. She has conducted several research projects in the field of gender studies, cultural processes, and communication and information technologies. She is the author and editor of many books, is associate editor of the journal The Information Society and serves as referee for many outstanding journals. She is active at European level especially in COST networks and is the Italian representative in the COST Domain Committee (ISCH, Individuals, Societies, Cultures and Health). She is the co-chair with Richard Ling of the International Association "The Society for the Social Study of Mobile Communication" (SSSMC). Her works have been published in eleven languages: Bulgarian, Chinese, English, French, German, Italian, Japanese, Korean, Russian, Slovenian, and Spanish.

Contributors

Afifa El Bayed M.A., is a Research Assistant and Lecturer of communication studies at the University of Erfurt. The fields of her teaching and research interest comprise mediated interpersonal communication, communication with media including social robots, as well as social media and collaborative innovation and production within their context.

Serena Cangiano M.Sc., is a Ph.D., candidate at University IUAV of Venice and SUPSI. She teaches and carries out projects of applied research at the Laboratory of visual culture at SUPSI University of applied sciences and arts of southern Switzerland, in the field of open design, interactive art, and user experience design. She coordinates the Master of Advanced Studies in Interaction Design SUPSI and the activities of the FabLab Lugano (www.fablab.supsi.ch). She is member of the Open Design + Hardware group of the Open Knowledge Foundation. Her experience spans from interaction design to user experience design, innovation and open design studies, electronic prototyping, and digital fabrication.

Chung-Tai Cheng Ph.D., gained his doctorate in Sociology from the Peking University in China. He is currently Teaching Fellow in the Department of Applied Social Sciences, The Hong Kong Polytechnic University. His research interest is in the topic of the social and normative implications of Information Communication Technology (ICT) use in contemporary China. His presentation 'Imagined performativity: the great virtue of cyberspace in contemporary Chinese workers' social lives' won the Best Presentation Award in COST Action 298 Conference: the Good, the Bad, and the Challenging in the year 2009. He has published several journal articles and book chapters and presented conference papers, mainly focusing on social and cultural impacts of ICTs.

Elizabeth Thomas Crocker is a Ph.D., student in Anthropology and a graduate research assistant in the Division of Emerging Media Studies at Boston University. She is interested in examining how emerging media is redefining experiences of religion, visual communication, togetherness, and being in the world. Recently, she has also explored concepts of self with relation to possession and the other.

Elda Danese Ph.D., teaches a graduate course of fashion design at the IUAV University in Venice, Italy. She is interested in contemporary visual cultures, in fashion, and fabric production. Danese has both practical experiences in curation and published scientific studies on contemporary textile and fashion.

Federico de Luca Ph.D., works as Statistician at the International Energy Agency (IEA/OECD), Paris. He has a Ph.D. in Social Statistics obtained from the Division of Social Statistics and Demography at the University of Southampton. Previously, he has worked as a research assistant for the universities of Chicago, Padua, Venice, and Southampton. His research interests also include the ways to evaluate professionals' knowledge and on how this might impact on policymakers' decisions.

Davide Fornari Ph.D., holds a doctoral degree in Design Sciences from University IUAV of Venice and is a tenured Teacher Researcher at the Laboratory of visual culture of SUPSI University of applied sciences and arts of Southern Switzerland, in Lugano, where he teaches History of graphic design and coordinates the Master of advanced studies in Interaction Design. He edited the Italian translation of Heinrich Wölfflin's Prolegomena to a Psychology of Architecture (Milan 2011) and the collection of essays Estetiche del camouflage. In 2012 he authored the essay Il volto come interfaccia ("Face as Interface") with a grant from Swiss National Science Foundation. He is a member of the permanent observatory of ADI (industrial design association) and of the editorial board of the magazine "Progetto grafico".

Daniel Halpern is Assistant Professor in the School of Communications at Pontificia Universidad Catolica de Chile (Catholic University of Chile) and head of the think tank TrenDigital, where he studies, teaches, and does consulting work on social media and strategic communication. His research focuses on methods and approaches for understanding the social consequences of communication technologies. He has published several papers on the role of social media in shaping interpersonal relationships, mediating social aspects such as civic participation, and affecting processes in learning with technologies.

Joachim R. Höflich Ph.D., is a Professor and researcher of communication studies at the University of Erfurt. The fields of his teaching and research expertise include media use and effects, media change, media integration, mediated interpersonal communication, mobile communication, and communication and relationships with media such as social robots.

Timo Kaerlein is a Research Fellow at the Department of Media Studies at the University of Paderborn. His research interests include media theory, digital culture, personal/mobile media, and social robotics. In 2012 he visited the Intelligent Robotics and Communication Laboratories at the Advanced Telecommunications Research Institute International (ATR) in Kyoto, Japan as a Cooperative Researcher. In his doctoral research he strives for an understanding of emergent relationships between human body and technological artifacts ensuing with the proliferation of mobile networked computers like smartphones. Recent thematically relevant publications: Kaerlein (2014) The Social Robot as Fetish? Conceptual Affordances and Risks of Neo-Animistic Theory. International Journal of Social Robotics, Online First, 23 Dec 2014, doi 10.1007/s12369-014-0275-6; Kaerlein T. (2012) Presence in a Pocket. Phantasms of Immediacy in Japanese Mobile Telepresence Robotics. communication +1(1).

James E. Katz Ph.D., is the Feld Professor of Emerging Media at Boston University's College of Communication. In addition, he directs its Division of Emerging Media Studies. The Division addresses the process of how new media technologies are created and introduced to users, the effects they have on users, and how technologies and the content they produce are moulded, co-constructed, and re-constructed by users. Dr. Katz joined BU from Rutgers University where he held the title of Board of Governors Professor of Communication, the highest honor the University can bestow on a member of its faculty. During his time at Rutgers, Katz served two terms as chair of the Department of Communication and also directed the Center for Mobile Communication Studies, which he founded in 2004. Earlier in his career, Dr. Katz headed the social science research unit at Bell Communications Research, which also honored him with the title of Distinguished Member of Staff. Dr. Katz has devoted his career to analyzing the uses and social consequences of emerging communication technologies, especially the Internet and telephone. He explores how they affect social interaction and what their uses reveal about human nature and organizations and was among the first to demonstrate their pro-social uses. He also seeks to understand what the future holds in terms of society and communication technologies and works with others to explore ways in which society can best prepare itself to make the optimal use of new developments.

Pui-lam Law (Patrick) Ph.D., received his doctoral degree in Sociology from The University of New South Wales, Australia, and is currently Assistant Professor in the Department of Applied Social Sciences of The Hong Kong Polytechnic University. His research interest is on modernity and social development in China. He co-authored (2004) Marriage, Gender, and Sex in a Contemporary Chinese Village, co-edited (2006) New Technologies in Global Societies, edited (2012) New Connectivities in China: Virtual, Actual and Local Interactions, and was guest editor

(2008) of special issues on ICTs and China and on ICTs and Migrant Workers in Contemporary China for Knowledge, Technology and Policy 21(1–2). Recently, he is working on mobile media and political participation.

Antonio B. Moniz Ph.D., Professor at Universidade Nova Lisboa, Portugal, coordinator of the Ph.D. programme on Technology Assessment, and researcher at Karlsruhe Institute of Technology, Germany. He is Portuguese representative at the TC9 "Computers and Society" of International Federation of Information Processing, since 2006. Coordinated several scientific research projects—on the social implications of robotics in manufacturing industry, and on intuitive interaction between humans and industrial robot systems.

Mauro Sarrica Ph.D., earned his doctoral degree in Personality and Social Psychology. He is a Researcher at the Department of Communication and Social Research, Sapienza, University of Rome, Italy. His main interests are social construction of knowledge, stability and change of social belied, and peace psychology. He has been involved in international and interdisciplinary research on Internet and communication technologies. In these fields he has studied the social representations of ICTs, the impact of the Internet on journalists' practices and the use of online forums to motivate and coordinate collective actions.

Chapter 1
Introduction: Situating the Human in Social Robots

Sakari Taipale, Jane Vincent, Bartolomeo Sapio, Giuseppe Lugano and Leopoldina Fortunati

Traditionally the social has been considered as a characteristic of human beings, not of inanimate machines. At the same time, each technological device can be considered social born out of a complex process of invention, implementation, distribution and domestication by users (Hirsch and Silverstone 2004; Lasen 2013). Since recent technical developments have made possible rather detailed technical mimicking of human beings and their social features, and incorporating them in silicon chips, there is a pronounced need to understand to what extent the humanness can be implanted in social robots. This is also an occasion to think over and discuss what the human is when considered in this context of social robots. With this book we tackle what can be considered as a social robot, which in fact is a paradoxical term, from a social, cultural and humanistic perspective.

S. Taipale (✉)
University of Jyväskylä, Jyväskylä, Finland
e-mail: sakari.taipale@jyu.fi

J. Vincent
University of Surrey, Surrey, UK
e-mail: j.vincent@surrey.ac.uk

B. Sapio
Fondazione Ugo Bordoni, Rome, Italy
e-mail: bart@fub.it

G. Lugano
COST, Brussels, Belgium
e-mail: Giuseppe.Lugano@cost.eu

L. Fortunati
University of Udine, Udine, Italy
e-mail: fortunati.deluca@tin.it

J. Vincent et al. (eds.), *Social Robots from a Human Perspective*,
DOI 10.1007/978-3-319-15672-9_1

Contrary to the view of hard scientists who say "[N]ever ask a roboticist what a robot is. The answer changes too quickly" (Nourbakhsh 2013, pp 14–15), this book asks what a social robot is and what should we think about social robots in different areas of everyday life? Viewed from the perspective of social sciences and humanities, current social robotics research is taking its first steps. Until now, it has mainly followed the lines of human–computer interaction (HCI) studies. Despite their many strengths, HCI studies, when not solicited and inspired by social and human sciences as in the present case, can too easily confine technological artefacts to a perspective that privileges the machine in the human–technology interactions (Shaw-Garlock 2011). Hence, HCI studies provide an insufficient research window on the human side of the technological artefact. In this respect, social robots are not an exception. From a humanistic and social science perspective, that puts the human being at the centre of the analysis, humans become human because they are social and only individuals as social, political and cultural human beings can work as a model for the social that is now being embedded in robots. The social is still mainly defined by our experiences of human–human social interactions, although the interaction among robots and other intelligent machines is increasing.

In many ways, social robotics research has been technologically determined because engineering solutions, robotic design and technological exploration have dominated both the design of the robot and the studies on robots (e.g. Breazeal 2004; Nourbakhsh 2013). As with any scientific programme, a particular problem can be identified when it has an engineering/technological solution; consequently, often the problem is solved from a machine perspective. Instead of following this approach, which starts from the technology, this volume focuses on the human side of the human–technology interaction, where humans play a role as social, political and cultural actors. This book delivers new knowledge to reflect on what kind of society we want to build and live in and this includes the robots we build and use that also tell us about the model of our society (Fortunati 2013). Exploring social robots from our perspective means that we argue there is the need to start, not from the problems that can be solved, but from the social practices and how these practices interact with any technological artefact, social robots included. Human needs, senses, emotions, desires and attitudes all foreground the discourse.

Looking at robots from an insider social, cultural and humanistic perspective does not mean that we reject the benefits and outcomes of social robotics technology. Rather, the book maintains that there is the need to include human sciences and the probable users inside the design of social robots. Moreover, the book explores how some of these devices and tools have come to be absorbed into the everyday lives of humans and what are the limits, problems and advantages of such integration. Although probably many social robots were designed from a technologically determinist way the users have found ways to domesticate and engage in interaction.

In many ways, this book is about the social shaping of social robots, and about the adoption and integration of robots, machines and technologies by technologically more and more sophisticated humans. Often the machines people decided to adopt made their lives better and helped them solve ad hoc problems, such as by using a set of complicated and wire strewn connectivity linking up devices and using them for purposes for which they were not designed (Ely et al. 2011). People use machines in real time and when they want to, adapting a little to accommodate the inevitable limitations and, at times, unpredictability of the machine as well as their own human frailties.

The designers and producers of social robots have tried to overcome the mismatch between machine-to-machine and human-to-human interaction in various ways, making their machines look uncannily like humans (e.g. Bartneck et al. 2009), like science fiction film (or movie and cartoons) characters (*I robot*, Asimov 1950; The Terminator 1984; Futurama), or by focusing on the dexterity of the device or prosthesis (e.g. drummer arm Georgia Tech[1]; healthcare robots[2]—Scuola Superiore di Sant'Anna di Pisa) rather than on their charm or look. The machine and its operator is often the more powerful party in the relationship with the user who must follow the directions or respond to the capabilities and performance of the machine. Designers of social robots are constantly looking for ways to make their solutions more suitable to meet people's needs, more responsive and reactive to human behaviours, and able to 'think', but the speed, dexterity and most of all the unpredictability, frailty and vagaries of human behaviours have proved to be very difficult to conquer.

Understanding more about human's everyday life practices is foundational to usefully and practically incorporating social robotics into daily routines. This is also the centre of the debate in this volume. Viewed from the human perspective, the volume explores how humans have adopted and accommodated machines into their lives.

The standpoint adopted in the book may also be explained by dividing the notion of social robots into two parts: social and robot. In engineer-driven robotics, the human body is typically taken as a starting point when the aim is to develop sociable robots. It is only when the physical (often metallic or plastic) body of the robot is sufficiently developed and ready for test use that softer and smart technologies making the robot social are considered. In that moment, designers try to embed them in the hardware. These social technologies include many information and communication technologies (ICTs) and applications that are already on the consumer market as well as artificial intelligence. Often less attention is paid to the look and dressing of the robots, which are also important elements making the robot a more cultural part of the social fabric (Fortunati 2014).

Let us consider now social robots that are targeted at the domestic sphere. People who are expected to accept them are not used to having industrial-looking robotic devices in their homes (the majority do not have them in their work places

[1]http://www.news.gatech.edu/2014/03/05/robotic-prosthesis-turns-drummer-three-armed-cyborg.

[2]http://www.robot-era.eu/robotera/index.php?pagina=pagine_personalizzate&blocco=93&id=96.

either), thus perhaps an alternative approach may be more successful. A large proportion of people in the industrialised North, and also in the global South, are already familiar with many social ICTs (e.g. mobile phones, the Internet and tablet computers). Hence, it may be well-justified to also analyse these widely established technologies (see, Vincent and Cheng in this volume) with the aim of understanding how they are, step by step, being automatized and then roboticised by incorporating into them new sophisticated features, applications (e.g. Apple's voice-driven personal assistant Siri) and even physical extensions, which make them social robots (e.g. telepresence robots like Double[3] which combines an iPad with a movable base). By taking the everyday life of ordinary people as a starting point, we may succeed painting a realistic picture of how social robots will become part of our homes and other private spaces in the foreseeable future.

This book presents an anthropocentric perspective of the social robot by bringing together studies from three continents: Europe, Asia and North America. We aim to deliver a balanced contribution of empirical studies (both qualitative and quantitative) and theoretical accounts, in order to widely examine the personal, social and cultural dimension of social robots. The book is divided in three thematic sections. The first of the three themes is *Perceptions and Attitudes to Social Robots*. The role of social robots in society is still relatively unexplored including the understanding of the reasons for their creation, the actions they perform and their future roles in society. This section brings together chapters that investigate perceptions of and attitudes towards social robots by using surveys and research materials collected both nationally and cross-nationally, and by examining country specific data from both quantitative and qualitative perspectives.

The shift of robots from the field of economic production to the sphere of social reproduction where people recuperate their physical, affective and mental resources is discussed in the chapter written by *Sakari Taipale, Federico de Luca, Mauro Sarrica and Leopoldina Fortunati*. By conducting a secondary analysis of large-structured survey data collected in Europe, Taipale et al. show that the general attitude among Europeans towards robots is positive, anticipating their further penetration in the markets. Even if Europeans are quite unwilling to have more robots, especially in the sphere of the social reproduction, the results of the study disclose some unexpected features in people's opinions. Contrary to common beliefs, pensioners are among the most accommodating of Europeans concerning the use of robots in the health care sector, elderly, child and disabled care, and education. The study by *James Katz, Daniel Halpern and Elizabeth Thomas Crocker* deals with the perception and acceptance of robots among US college students. They contribute to the current knowledge by showing that the human-likeness of robots is positively associated with students' willingness to accept them. Perhaps more interestingly their results clearly indicate that the acceptance of social robots may be successfully promoted by engaging people first in the use

[3]http://www.doublerobotics.com/.

of online communities. Katz et al. show that it is the previous engagement in the online platforms of social communication through avatars and a sense of belonging to an online community that predict students' readiness to accept robots. This study is a signpost indicating that rather than the engineer-driven mimicking of human-likeness, it is more people's previous experience in engaging in virtual worlds that paves the way for social robots.

The perception and acceptance of social robots is also studied by *Joachim Höflich and Afifa El Bayed*, who investigate people's stance towards robots in Germany. Their study shows how the perception and social representation of robots is powerfully influenced by popular culture and media. Further, they show with a small experiment that although people, who try to teach a robot to perform simple tasks associate much humanity to robots, this humanity also easily disappears when the robot fails to perform certain tasks.

The second part of the book moves from the perceptions and attitudes towards social robots to *Human Interaction with Social Robots.* This section presents four chapters the first two of which explore anthropomorphism in human–robot interaction, especially in relation to the process of robot design. *Patrick Law's* examination of the transformation of wheelchairs into social robots in Hong Kong highlights that the user experience of being made to perform daily tasks within the constraints of a machine can lead to their rejection of it. The problems caused by machines must also be taken into consideration while introducing robots into the workplace and *António B. Moniz* argues that even in the productive sector many problems related to robots are not technical but are framed by social aspects. He maintains that increasingly advanced and sociable robots do not solely make people's work easier and safer, but present new job demands and psychological strain for human operators. *Timo Kaerlein* also addresses the design of robots underlining the need for reducing the complexity of robots by making them human-like only to a very limited extent. Drawing on the cybernetic theoretical perspective, he brings to the fore the risk that robotics may also promote dehumanisation and explores ethical considerations in this regard. His examination of new robot devices that can be used to express feelings in communications contrasts with Vincent's later chapter. The final chapter of this section is authored by *Davide Fornani and Serena Cangiano* and it deals with some examples of participatory design of robots. Drawing from several open-source robot projects, which engage a wider group of users in the design and development of robots, the authors argue that in open-source development there is a higher possibility that human needs are recognised and incorporated in robots than in engineer-driven development. The authors conclude by proposing some guidelines for designers that should help them design robots that better meet human needs.

The third theme of the book examines the role of *Social Robots in Everyday Life* by way of three chapters that present qualitative analyses on human/social robot relationships. What is common to these chapters is that they all elaborate possible pathways to a more widespread adoption of advanced social robots by investigating automatized and robotised features that are already embedded in ordinary information and communication technologies, and in new wearable technologies.

Jane Vincent takes the mobile phone as an example of an ordinary device that in close dependent interaction with its user is developing into an "emotionalised social-robot" which effectively evokes emotions in the users. Her chapter reflects one of the most social, humanistic and social scientific approaches to social robots so far and challenges the normative approach to social robots that are expected to be involved in the field of social reproduction by arguing that immaterial responses and communicative actions to a user's behaviour from a robotic device (e.g. stirring up emotions, predictive text-input learning a user's vocabulary) are equally and sometimes even more important than physical actions that robots can perform (e.g. the manipulation of physical items). *Chung-tai Cheng*'s contribution considers the mobile phone—and more specifically a mobile phone based home-school communication system—as a quasi-robot that indicates a shift towards a more automatized and roboticised educational system in China. While this approach to social interaction may serve the efficient management of Chinese mass education, at the same time these processes may strengthen negative aspects of competitiveness between families and the suppression of individuality in schools. In the last chapter, *Elda Danese* takes the close relationship between new technology and fashion as her starting point while analysing the possible futures of wearable social robotics. Fashion typically heralds wider societal trends and changes and Danese takes us through the scope of a broad spectrum of projects examining wearable technologies and body-related devices, including artificially intelligent systems that react to changes in environments. In her chapter, fashion becomes portrayed as a feasible way of introducing social robotics to our day-to-day life.

The book ends with a concluding chapter from the editors, *Jane Vincent, Sakari Taipale, Bartolomeo Sapio, Giuseppe Lugano and Leopoldina Fortunati.* Here the editors weigh the pros and cons of the human standpoint to robotics adopted and applied in the volume. They assert that the proposed human perspective helps demystify the role of robots in contemporary (and future) societies. They also remind the readers that the chapters of the book show that people's attitude towards robots is chiefly positive, despite some common fears and threats that often label public discourses. According to the editors, the book also gives reasons to question some deep-rooted ideas in robot research. One of them relates to the superiority of robotic hardware (manipulators, artificial limbs etc.) and robotic software as the determinants of what make the robot social. The other concerns the superiority of a human face to a screen as a user interface. As people have accommodated screen-equipped communication and media devices, which are also increasingly intuitive to use, the screen could provide a genuine gateway for bringing social robots into homes. The conclusion—and with it the whole book—is ended with a short overview of the limitations and future research needs.

References

Bartneck C, Kanda T, Ishiguro H, Hagita N (2009) My robotic doppelgänger—a critical look at the uncanny valley. In: Proceedings of the 18th IEEE International Symposium on Robot and Human Interactive Communication, IEEE, Piscataway, 269–276

Breazeal C (2004) Designing sociable robots. MIT Press, Cambridge

Ely P, Frohlich D, Green N (2011) Uncertainty, upheavals and upgrades: Digital-DIY during life-change. In: Pierson J, Mante-Meijer E, Loos E (eds) New media technologies and user empowerment. Peter Lang Publishing, Berlin, pp 163–180

Fortunati L (2013) Afterword: robot conceptualizations between continuity and innovation. Intervalla 1:116–129

Fortunati L (2014) Robots approaching the human body: roboticizing fashion and design. Presentation at the workshop A sociological, psychological and educational perspective on social robots, Biorobotics Institute of the Scuola Superiore Sant'Anna, Pisa, Italy

Hirsch E, Silverstone R (eds) (2004) Consuming technologies: media and information in domestic spaces paperback, New edn. Routledge, London

Lasen A (2013) Digital inscriptions and loss of embarrassment: some thoughts about the technological mediations of affectivity. Intervalla 1:85–100

Nourbakhsh IR (2013) Robot futures. The MIT Press, Cambridge

Shaw-Garlock G (2011) Loving machines: theorizing human and sociable-technology interaction. In: Lamers MH, Verbeek FJ (eds) Human-robot personal relationships. Springer, Heidelberg, pp 1–10

References

Part I
Perceptions and Attitudes to Social Robots

Chapter 2
Robot Shift from Industrial Production to Social Reproduction

Sakari Taipale, Federico de Luca, Mauro Sarrica and Leopoldina Fortunati

This chapter analyses people's attitudes towards the use of robots in the different domains of life and, specifically, in the domain of social reproduction. The analysis is based on Eurobarometer 382 "Public Attitudes towards Robots" data ($N = 26{,}751$), which was carried out among EU citizens aged 15 and over in 27 member states in 2012. The results of the study show that on average European perceptions of robots are positive and permissive. The life domains in which robots have already been used for a long time (e.g. space exploration, manufacturing, military and security business, search and rescue work) turn out to be the most popular areas for the further penetration of robots. The least preferred life domains are those, which address the core functions of social reproduction (e.g. care of children, elderly people and the disabled, education, leisure). With a series of ordinal logistic regression analyses, we outline the socio-demographic factors that are associated with the willingness to have more robots in the various fields of social production. Pensioner's supportive attitude towards the use of robots in health care and educational activities is highlighted.

S. Taipale (✉)
University of Jyväskylä, Jyväskylä, Finland
e-mail: sakari.taipale@jyu.fi

F. de Luca
International Energy Agency, Paris, France
e-mail: beckowen86@yahoo.it

M. Sarrica
University of Rome, Rome, Italy
e-mail: mauro.sarrica@uniroma1.it

L. Fortunati
University of Udine, Udine, Italy
e-mail: fortunati.deluca@tin.it

J. Vincent et al. (eds.), *Social Robots from a Human Perspective*,
DOI 10.1007/978-3-319-15672-9_2

2.1 Introduction

The industrialised world is experiencing a shift from industrial to social robotics. In terms of social politics, this is an epoch-making change; while automation and industrial robots first and foremost have affected the demand for mechanical human work and working conditions in the industrial work (e.g. Zuboff 1988; Rifkin 1995), social robots are being designed to deal with human care, health, domestic tasks, entertainment and various other forms of immaterial and material tasks which aim to renew human capacities. It is these typically non-monetized domestic tasks that are gathered under the concept of social reproduction (Fortunati 1995). Consequently, for the first time we have to think about the possible consequences of robots, not only for industrial production but also for social reproduction. This shift leads us to study and discuss the relationship between social robots and the policies that are framed to deal with the problems of social reproduction.

In this chapter, we analyse people's attitudes towards the use of robots in the different domains of life and, specifically, in the domain of social reproduction (e.g. care, domestic tasks, education, leisure and health). To this end, three research questions are defined. First, what is the overall attitude towards the use of robots in Europe? Second, in which domains of life are Europeans most willing to see more robots in the future? Third, who is most likely to accept robots in the field of social reproduction in Europe? The types of robots which are within the scope of this research are human-like and instrument-like robots.

The chapter begins with a description of robots as the targets of social research. We argue that while industrial robots were involved with a relationship with strong social groups (like organised labour force), social robots deal increasingly with the weak social groups, like children, disabled and the old people, which call for special attention from social researchers. Where the industrial sector means guaranteed and formalised salaries, the strong presence of unions, and highly formalised work contracts, the reproduction sphere means unwaged work (domestic), lack of support from unions, and limited formalised negotiation, because it is mainly considered a "private" sphere. We will then undertake a secondary analysis of Eurobarometer 382 "Public Attitudes towards Robots" data ($N = 26{,}751$), which was carried out among EU citizens aged 15 and over in 27 member states in 2012. With this analysis, we will answer our three research questions, and we will then conclude the chapter by discussing the types of social policies that are needed in the era of robot-mediated and robot-assisted social reproduction.

2.2 What Changes When Robots Enter the Social Sphere?

The societal role of robots has changed profoundly with the shift from industrial to domestic robots. There were three main reasons for using robots in industrial sectors. Robots were introduced to take care of dangerous and/or repetitive jobs, to save human labour and human lives. They were employed to improve the quality of products (e.g. gluing, spraying, testing, gauging), and production processes (e.g. assembling). Finally, robots were superior to human labour in terms of guaranteeing the regularity of the work, including the fact that they do not go on strike. The introduction of domestic and assistive robots has enabled us to refine and update our understanding of robots beyond that of being machines that perform dangerous, dirty and monotonous work (Dautenhahn et al. 2005; Kim and Mutlu 2014).

The reasons for using robots in the domestic sphere are different from those for industrial production. In the domestic sphere, in fact, the products and processes with which social robots are expected to be involved are not dangerous for humans. Similarly, their ability to maintain the regularity and speed of work are less appreciated properties in this context. On the contrary, social robots are expected to "make good" and take care of people, work which presents serious problems of formalisation. Thus, flexibility, adaptability to personal needs and the ability of complex reasoning are more relevant properties for social robots. It can also be posited that the increasing interest in the use of robots for social reproduction might be time related. Women, especially, have increased the time spent on paid-work between 1970 and 2000, which was maybe accompanied by a decrease in their supply of unpaid domestic work (Gimenez-Nadal and Sevilla 2012). There is also evidence from some countries that the number of people helping grown-up family members has dropped between the 1990 to the 2000 (Pääkkönen and Hanifi 2011). At the same time, the spatial and temporal dispersion of families makes family-based care increasingly difficult to arrange. It is against this backdrop that social robots can be seen as a way to compensate for this lack of time.

It is equally important to realise that social robots face very different kinds of counter forces than did industrial robots. While industrial robots function in semi-public spaces (e.g. factories) which are usually controlled by managers and regulated by workplace policies and labour laws, social robots enter the private sphere of life, which unfolds in people's own homes or in nursing homes. Therefore, the questions of privacy, intimacy and affects, as well as the issue of care labour, come to the fore. What kind of shared rules and public policies do we need to make sure that robots are introduced and used in the sphere of social reproduction in a proper manner? Second, while industrial robots faced the relatively well-organised and strong labour movement that defended employee rights, today's trade unions are somewhat weaker in the face of the global competition over the waged and, particularly, the unwaged labour force. Actually, the protecting hand of trade unions has never reached unpaid domestic work. If a ghost in the industrial sector is the fear of losing a job due to the introduction of robots in certain sectors, in the domestic sphere this fear does not affect housewives or househusbands who work without any economic retribution. This fear might, however, affect waged-workers and professionals in the health care, education, social care and entertainment sectors. Thus, in the sphere of reproduction social robots do not need to fight against labour movements and trade unions, but they do need to win over the confidence of individuals and families, with whom they are supposed to work. Finally, earlier robots were designed for industrial work, which mostly employed male labour.

Social robots, conversely, are designed to deal with social reproduction, such as taking care of children, the elderly, the disabled and the ill, tasks which have traditionally been carried out mainly by women (Boyer 2004; Sparrow and Sparrow 2006). The importance of this kind of reproductive work has never been truly recognised by policy makers, as reflected, for example, in the fact that reproductive work is not included in the gross domestic product (GDP). All in all, social robots are mainly expected to deal with weak social groups, something which makes it important to study this new generation of robots from the perspective of social and policy studies. In addition, social robots work as a kind of boundary object, marking the line between low-tech home technologies and high-tech industry, masculine devices (robot) and feminine work (e.g. social work, health care), commodity production (robots for mass-production) and individual reproduction (e.g. personalised assistive robots).

This shift from production to reproduction was preceded and supported by the introduction and development of a series of automated processes of social behaviours, which are strongly creating an acceptance of the robotification of society. We refer here to the processes of automation which affect the areas of taste (Barile and Sugiyama 2015), communication (Bakardjieva forthcoming; Baron forthcoming), information (Vámos 2009), education and play (Fortunati 2015). These processes can be considered as forms of proto-robotification of the immaterial sphere in society, made necessary by digital technologies that allow an incredibly large number of people to express their opinions and tastes, to buy and consume goods (e.g. Shirky 2008).

2.3 From Individuals to Societies and Beyond

In spite of the fact that robots are now entering the home, the analytic level of social robot studies has been quite limited. It is somewhat paradoxical that while the role of automatisation and industrial robots has been generally considered, perceived and analysed as a societal challenge that transforms the labour market and work processes (e.g. Rifkin 1995; Frey and Osborne 2013), social robots have been so far studied largely by analysing the perception and behaviours of only small groups of individuals (Meister 2014, p. 113). More precisely, roboticists, designers, sociologists and psychologists have mostly investigated individual and situational factors, such as the perceived human-likeness of a robot (Halpern and Katz 2014) or the physical distance between a robot and human (e.g. Kim and Mutlu 2014), that might affect the acceptance and perception of robots (also, Meister 2014). We argue that it is equally—if not sometimes more—important to examine the life domains in which people at national and supranational levels are most willing to live and work with social robots (see also Enz et al. 2011; Takayama et al. 2008). Identifying the wider attitudes and perceptions, as well as the type of people behind them, contributes to a better understanding of robots. As social robots enter the reproduction sphere, including elderly people's homes, children' rooms and kindergartens, it is crucial to understand which of these areas are receptive to robots.

To solve the large and complex social problems of contemporary societies, such as those related to the use of robots in social reproduction, both scientific knowledge and the opinions of citizens must be included in the decision-making process. While new technical solutions are certainly needed in this endeavour, it is equally important to understand how economic, social, political and cultural aspects of life shape the contemporary attitudes and opinions of citizens towards robots.

2.4 Data and Measurements

This chapter is based on a secondary analysis of Eurobarometer 382 "Public Attitudes towards Robots" data ($N = 26{,}751$).[1] This survey was carried out among EU citizens aged 15 and over in 27 member states in 2012. Respondents were interviewed face-to-face at home in their mother tongue (Eurobarometer 2012). To our knowledge, this survey is the most extensive so far, although not issue-free, attempting to gauge public opinion and citizens' attitudes towards robots. It covers a wide area of robotic applications and includes a good set of attitude measures. At the same time, however, it provided respondents with a rather narrow image of what robots were by presenting only two pictures of robots (Eurobarometer 2012, p. 4). The characteristics of the sample are reported in Table 2.1.

In order to investigate European attitudes towards robots in the shift from the industrial to the domestic sector, we analysed the answers given to the following two questions: "In which areas do you think robots should be used as a priority?" and "In which areas do you think that the use of robots should be banned?" A maximum of three answers per question was allowed. For both questions, the following domains of life were supplied as answer choices: Manufacturing; healthcare; leisure; domestic use (such as cleaning); military and security; search and rescue; education; care of children, elderly, and the disabled; space exploration; agriculture; transport/logistics. Respondents could also choose "other domains," "none of them" or reply "I don't know." Based on these two original questions, we created a new measure to indicate the *overall attitude towards robots* (range = −3 to 3, mean = 0.679, standard deviation = 1.199). This variable summarises the answers regarding the use and banning of robots in the above-mentioned fields. The answers to the first questions were given a positive sign, while those to the second were given a negative sign, and thus this new variable ranges between −3 and +3. We then created ordinal variables indicating the willingness to see robots in the various domains. The values of these *domains-specific attitude* measures vary between −1 (ban), 0 (indifference) and 1 (use as a priority).

[1]European Commission, Brussels: Eurobarometer 77.1 February–March 2012. TNS OPINION & SOCIAL, Brussels [Producer]; GESIS, Cologne [Publisher]: ZA5597, dataset version 2.0, doi: 10.4232/1.11481.

Table 2.1 Description of the sample (non-weighted)

	%	*N*		%	*N*
Gender			*Children*		
• Men	48.3	12,928	• Family with children	40.3	10,786
• Women	51.7	13,823	• Family without children	58.3	15,604
Total	100.0	26,751	• No answer	1.4	361
			Total	100.0	26,751
Age			*Degree of urbanisation*		
• 15–17	4.9	1,322	• Rural area/village	34.0	9,095
• 18–24	9.4	2,516	• Small/medium-sized town	40.1	10,719
• 25–44	32.7	8,744	• Large town/city	25.9	6,916
• 45–64	32.2	8,623	• No answer	0.1	21
• 65+	20.7	5,545	Total	100.0	26,751
Total	100.0	26,751			
Education			*Country*		
• Low education	20.1	5,365	• Austria	3.9	1,031
• Medium education	42.4	11,348	• Belgium	3.9	1,051
• High education	30.8	8,227	• Bulgaria	3.8	1,006
• No Answer	6.8	1,812	• Cyprus	1.9	506
Total	100.0	26,751	• Czech Republic	3.7	1,003
			• Denmark	3.8	1,019
Years of education			• Estonia	3.7	1,000
• 15 or less	21.1	5,570	• Finland	3.7	1,003
• 16–19	47.6	12,540	• France	4.0	1,059
• 20 or more	31.2	8,227	• Germany	5.8	1,552
• No answer	1.6	415	• Greece	3.7	999
Total	100.0	26,751	• Hungary	3.8	1,021
Activity			• Ireland	3.8	1,008
• Worker	49.2	13,163	• Italy	3.9	1,036
• Housewife/-husband	7.9	2,107	• Latvia	3.8	1,024
• Unemployed	8.5	2,280	• Lithuania	3.8	1,021
• Pensioner	25.3	6,761	• Luxembourg	1.9	501
• Student	9.1	2,441	• Malta	1.9	500
Total	100.0	26,751	• Netherlands	3.8	1014
			• Poland	3.7	1,000
			• Portugal	3.8	1,009
Social class			• Romania	3.8	1,020
• Low	20.5	5,487	• Slovakia	3.7	1,000
• Medium	50.7	13,552	• Slovenia	3.8	1,017
• Medium-high	26.2	7,002	• Spain	3.8	1,004
• No answer	2.7	710	• Sweden	3.8	1,016
Total	100.0	26,751	• United Kingdom	5.0	1,331

(continued)

Table 2.1 (continued)

	%	N		%	N
Income			Total	100.0	26,751
• Low	9.9	2,641			
• Medium	25.4	6,789			
• Medium-high	62.0	16,588			
• No answer	2.7	732			
Total	100.0	26,751			
Family					
• Couple with children	32.2	8,616			
• Couple without children	29.4	7,860			
• Single	28.9	7,744			
• One parent and children	6.4	1,714			
• Mixed families	1.7	456			
• No answer	1.4	361			
Total	100.0	26,751			

2.5 Results

Eurobarometer (2012, p. 6) report on the analysed data concludes that, among all EU citizens, 70 % have either a positive or very positive view of robots. To further investigate Europe's general stand on robots, we analysed the overall attitude indicator. Figure 2.1 shows that a large number of respondents, 41 %, provided the same number of answers to both questions inquiring about the areas of life in which robots should or should not be used. Almost 50 % of respondents chose more positive responses than negative ones. Only 10 % of respondents selected more domains where robots should be banned than those where they should be used. In response to the first research question, it results that on average European perceptions of robots are positive and permissive.

Regarding the second research question, Table 2.2 illustrates the life domains where Europeans would like to use robots as a priority. It turns out that life domains in which robots have already been used for a long time, such as space exploration and manufacturing, as well as life areas in which robots can clearly save human lives, such as military and security business and search and rescue work, are the most popular areas. After these four domains, the public support for robots drops quite dramatically.

Approximately 22 % of Europeans would like to see robots first and foremost in health care, 13 % in domestic activities, transportation and agriculture. Clearly, the least preferred life domains are those, which serve the core functions of social reproduction.

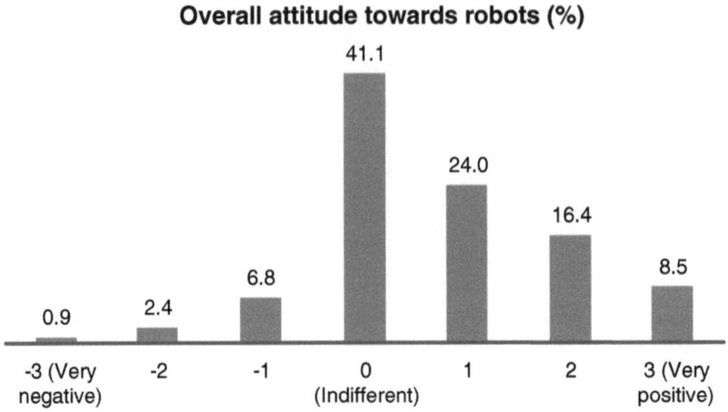

Fig. 2.1 Overall attitudes of Europeans towards robots (%)

Table 2.2 The domains of life in which robots are supported by Europeans (weighted)

Life domain	*n*	%
Space exploration	13,895	51.9
Manufacturing	13,282	49.7
Search and rescue	11,016	41.2
Military and security	10,937	40.9
Health care	6,007	22.5
Domestic activity	3,574	13.4
Transportation	2,962	11.1
Agriculture	2,813	10.5
Child/elderly/disabled care	947	3.5
Education	694	2.6
Leisure	670	2.5

Note Domains of social reproduction are in bold

These include care of children, elderly people and the disabled, education where the labour force of tomorrow is qualified and updated, and leisure, which refers to the time that people take for themselves to recover from work and their other duties.

As the shift from production to reproduction in robot use and research forms the basis of this study, we ran a series of ordinal logistic regressions in order to investigate the socio-demographic profile of the Europeans who are ready to see robots in social reproduction. Table 2.3 presents the results of these regression analyses considering health care, child/elderly/disabled care, domestic work, leisure and education, which enables us to answer our third research question.

Europeans who consider themselves as belonging to the highest social classes are more willing to see robots in health care than people who consider themselves in the lower social strata. Pensioners are more in favour of robots in health care than any other activity group, while single parent families are less receptive towards the use of robots in health care than the other family types. Willingness to have more robots in health care also increases with years of education and size of place of residence.

Table 2.3 Ordinal regression models for domain-specific attitudes

Independent variable (reference group)	Health care	Care of children, elderly, and the disabled	Domestic use	Leisure	Education
Social class (*lowest*)					
Medium	−0.065 (0.051)	–	–	–	–
High	0.170 ** (0.060)				
Activity (*Employee*)					
Housewife/househusband	−0.072 (0.083)	0.058 (0.084)	–	–	0.163 (0.085)
Unemployed	0.034 (0.073)	0.033 (0.777)			0.077 (0.778)
Pensioner	0.221*** (0.051)	0.188*** (0.048)			0.188*** (0.052)
Student	−0.090 (.072)	0.224** (0.081)			−0.066 (0.087)
Family (*couple with children*)					
Couple without children	−0.061 (0.051)	–	0.070 (0.062)	–	–
One parent and children	−0.126* (0.053)		0.101 (.066)		
Mixed families	−0.097 (.084)		0.206 (0.108)		
Single	−0.154 (0.122)		−0.290* (0.144)		
City (*Rural area/village*)					
Small or medium-sized town	0.020 (.046)	0.118**(0.050)	–	–	–
Large town/city	0.150** (0.049)	0.155** (0.052)			
Years of Education (15 *or less*)					
16–19	0.109* (0.055)	–	0.281*** (0.073)	−186** (0.068)	−0.066 (0.059)
20 or more	0.294*** (0.061)		0.377*** (0.078)	−0.066 (0.071)	−0.165* (0.065)

(continued)

Table 2.3 (continued)

Independent variable (reference group)	Health care	Care of children, elderly, and the disabled	Domestic use	Leisure	Education
Age (15–18)					
19–24	–	–	−0.369* (0.170)	–	–
25–44			−0.448** (0.157)		
45–64			−0.655*** (0.155)		
65+			−0.637*** (0.159)		
N	24,916	25,119	24,741	24,632	24,663

* = $p < .05$; ** = $p < .01$; *** = $p < .001$

Note Standard errors are in brackets. A positive coefficient indicates greater agreement with the use of robots. All models are controlled for the effect of country differences. Of other independent variables, only significant predictors were included

The use of robots for caring for children, the elderly and the disabled was predicted only by the respondents' main activity and the size of respondent's place of abode. In this respect, pensioners and students seem to be keenest to support the introduction of robots. It is particularly surprising to see that pensioners are significantly more likely to accept the use of robots in the care sectors than are the employed. People living in large towns and cities are more open to have more robots in the care sectors than people living in smaller town or villages.

Robots in the domestic activities seem to be mainly associated with three variables. First of all, singles are less likely than families with children to support the use of robots for domestic tasks. This makes sense as singles have comparatively fewer domestic duties than other kinds of family, and they would thus benefit less from domestic robots. Second, medium and long-term education is positively associated with the use of robots in domestic tasks. Third, the older the respondents are, the less likely they are to believe that there should be more robots taking care of domestic tasks.

In terms of supporting the introduction of robots to leisure activities, responses are instead particularly associated with the number of respondent's years of education. Those with from 16 to 19 years of education, which corresponds to a Bachelor's or Master's degree, are less likely than respondents with less education to support robots in leisure activities. Finally, people's willingness to see more robots in the education sector is mainly related to two factors: pensioners are more open to such robotic solutions than workers, while, conversely, the most educated people (with 20 or more years of education) are in this respect more critical than less educated.

2.6 Discussion and Conclusions

In this study we analysed European opinions on robots in general and, specifically, in the sectors of social reproduction. The analysis shows that a shift from their use in production to reproduction is not wanted by the majority of Europeans. Although the overall attitude towards robots is positive, permissive attitude towards the penetration of robots in the field of social reproduction in particular is still limited. A more careful investigation of this opposed area revealed some intriguing weak, but important signs for the future of social robotics.

First, against all prejudices, pensioners do not seem to be reluctant about the use of robots in social reproduction. On the contrary, they are significantly more supportive of the use of robots in health, care and educational activities than, for instance, workers. Thus, one may provocatively ask whether those who are concerned with the wider use of social robots might be their loved ones, who feel guilty about their inability to provide care, rather than the ageing pensioners themselves, who would benefit from assistive robots.

In terms of social policy, we must be careful not to neglect pensioners' own will by falling into stereotypic beliefs that older people are not ready for technological innovations. Some experimental studies also hint that elderly people accept assistive social robots surprisingly well (Cavallo et al. 2014), even if the acceptance of robotic help seems to depend on the type of task. While a robot is sometimes preferred over a human by elderly people in instrumental tasks such as changing a light bulb or doing laundry, humans are favoured over robots in personal tasks, such as care and leisure activities (Smarr et al. 2014). Although they are easily seen as a weak social group (Fortunati 2014), we must remember that today's pensioners, if compared to the elderly of the past, are wealthier, live longer and more technologically savvy. Consequently, they should not be seen as passive recipients of social robotic technology, but as active co-designers and selective consumers of future robot devices and applications.

Second, as so often in the history of new technologies, large cities seem to be the most favourable places for the introduction of robots into the health and care sectors. While in villages and smaller towns attitudes towards the care and health services seem to still be more families and community oriented, in the large cities people are used to relying more on external or technological aid. Robots replacing a non-family caregiver, or an unknown medical professional, or acting as mediators between a person and a family doctor, might not be considered such a big problem as a robot replacing a family carer. It is also interesting to note that for the use of robots in domestic tasks and less intimate fields of social reproduction, such as leisure and education, the size of the place of residence does not matter.

The answers of the respondents to this Eurobarometer survey, who represent the voice of European citizens, are very important. Even if people still fear that they might lose jobs because of robots, the use of robots in the production sector is currently not their main concern when compared to social and individual reproduction sectors. In conclusion, the main concerns relate to the use of robots in care, healthcare and education. These are the areas that involve many social interactions and much human contact, and which in the age of sociable robots are presented as the new and prominent environment to which robots should be introduced. The concern regarding fewer human contacts partly relates to a misbelief that the use of social robots always aims to replace human beings or human tasks, when in fact the most frequent aim of social robots is to support and enhance user abilities or assist human carers in heavy tasks (Sparrow and Sparrow 2006; Cavallo et al. 2011). This misbelief is very probably related to the idea that the same logic that applied to the production sphere, where the introduction of robots primarily aimed to replace workers with more efficient machines, would be exported to the field of social reproduction. On the contrary, the logic that shapes the reproduction sphere is different, if not opposite, because the needs to be met are different.

From a technical point of view, social robots could be designed and developed to serve almost every aspect of daily life. Moreover, a substantial contribution from robots and other novel technologies is especially needed to ease reproductive tasks in the domestic sphere, when the double burden of work and care on women in particular is heavy in current European societies. However, at the same time there is a resistance among Europeans towards the penetration of robots in the reproductive sphere. Taking public opinions into account in the decision making and trying to understand the reasons behind these opinions, increase the possibility that the design and implementation processes of social robots will be more suitable to European citizens' needs and not immediately rejected or prejudged.

References

Bakardjieva M (forthcoming) Rationalizing sociality: an unfinished script for socialbots. Inf Soc 31(3)
Barile N, Sugiyama S (2015) The automation of taste: a theoretical exploration of mobile ICTs and social robots in the context of music consumption. Int J Soc Robot
Baron N (forthcoming) Shall we talk? Conversing with humans and robots. Inf Soc 31(3)
Boyer K (2004) The robot in the kitchen: the cultural politics of care-work and the development of in-home assistive technology. Middle States Geogr 37:72–79
Cavallo F et al. (2011) Engineering in medicine and biology society, EMBC. Paper presented at the annual international conference of the IEEE, 30 Aug 2011–3 Sept 2011
Cavallo F et al (2014) Development of a socially believable multi-robot solution from town to home. Cogn Comput. doi:10.1007/s12559-014-9290-z
Dautenhahn K et al (2005) What is a robot companion—friend, assistant or butler? Proceedings of the IEEE/Robotics society of Japan international conference on intelligent robots and systems, IROS, vol 05, pp 1488–1493
Enz S et al (2011) The social role of robots in the future— explorative measurement of hopes and fears. Int J Soc Robot 3(3):263–271
Eurobarometer (2012) Public attitudes towards robots. Special eurobarometer 382: European commission. http://ec.europa.eu/public_opinion/archives/ebs/ebs_382_en.pdf. Accessed 3 Nov 2014
Fortunati L (1995) The arcane of reproduction. Housework, prostitution, labor and capital. Autonomedia, New York
Fortunati L (2014) Afterword: robot conceptualizations between continuity and innovation. Intervalla 1:116–129
Fortunati L (2015) Mobile media, play and social identities. In: Frissen V, Lammes S, de Lange M, de Mul J, Raessens J (eds) Playful identities: the ludification of digital media cultures. Amsterdam University Press, Amsterdam, pp 293–306
Frey CB, Osborne M (2013) The future of employment: how susceptible are jobs to computerisation? Oxford martin school working papers. http://www.oxfordmartin.ox.ac.uk/downloads/academic/The_Future_of_Employment.pdf. Accessed 2 Nov 2014
Gimenez-Nadal JI, Sevilla A (2012) Trends in time allocation: a cross-country analysis. Eur Econ Rev 56(6):1338–1359
Halpern D, Katz J (2014) Close but not stuck: understanding social distance in human-robot interaction through a computer mediation approach. Intervalla. http://www.fus.edu/intervalla/images/pdf/3_halpern_katz.pdf. Accessed 3 Nov 2014
Kim Y, Mutlu B (2014) How social distance shapes human–robot interaction. Int J Hum Comput Stud 72(12):783–795
Meister M (2014) When is a robot really social? An outline of the robot sociologicus. Sci Technol Innov Stud 10(1):107–134
Pääkkönen H, Hanifi R (2011) Ajankäytön muutokset 2000-luvulla. Stat Finl, Helsinki
Rifkin J (1995) The end of work: the decline of the global labor force and the dawn of the post-market era. G.P. Putnam's Son, New York
Shirky C (2008) Here comes everybody: the power of organizing without organizations. Penguing Books, New York

Smarr CA et al (2014) Domestic robots for older adults: attitudes, preferences, and potential. Int J Soc Robot 6(2):229–247

Sparrow R, Sparrow L (2006) In the hands of machines? The future of aged care. Mind Mach 16(2):141–161

Takayama L et al. (2008) Beyond dirty, dangerous and dull. In: Proceedings of the 3rd international conference on Human robot interaction—HRI '08, ACM Press, New York, USA, p 25

Vámos T (2009) Social, organization and individual impacts of automation. In: Nof SY (ed) Springer handbook of automation. Springer, Berlin

Zuboff S (1988) In the age of the smart machine: the future of work and power. Basic Books, New York

Chapter 3
In the Company of Robots: Views of Acceptability of Robots in Social Settings

James E. Katz, Daniel Halpern and Elizabeth Thomas Crocker

Society is moving toward a situation in which "socially intelligent robots" are part of human social life (Zhao 2006). Hence, it is important to explore how people experience, communicate with, and conceive of robots in their social and physical environments. Also important is to consider how people see the various roles for robots and how their comfort levels may vary depending on robotic functions and appearances.

3.1 Introduction and Literature Review

Since the mid-1980s, the themes human/robot interactions have continued to be explored not just from literary viewpoints but also from social scientific ones. Fortunati et al. (2003) delineated numerous ways that robots and humans have had meaningful interactions and become part of human conceptions. Katz (2003) showed that within numerous social domains there is a growing importance of electronic representations in social interaction. Studies done on human–robot interactions (HRI) also confirm that people tend to invest "human-ness" onto

J.E. Katz (✉)
Boston University's College of Communication, Boston, MA, USA
e-mail: profjameskatz@gmail.com

D. Halpern
Pontificia Universidad Catolica de Chile, Santiago, Chile
e-mail: dmhalper@uc.cl

E.T. Crocker
Boston University, Boston, MA, USA
e-mail: liminalphase@gmail.com

J. Vincent et al. (eds.), *Social Robots from a Human Perspective*,
DOI 10.1007/978-3-319-15672-9_3

artificial entities. For example, Friedman et al. (2003) analyzed conversations in discussion forums dedicated to Sony's robotic dog AIBO and found that 47 % of participants spoke about the its biological essences; 42 % of the forums' members spoke of AIBO as having intentional behavior; 38 % of the commentators believed it had feelings, and most incredibly 39 % spoke of AIBO as being capable of being raised, developing, and maturing.

The anthropomorphic qualities of robots lead human partners to treat humanoid social robots as real persons (Fong et al. 2002; Duffy 2003). Likewise, when robots are capable of natural language speech patterns and can demonstrate self-directed behaviors, human observers tend to attribute human-like qualities to the robots. This is because the robots invoke social-psychological processes able to affect human behavior in ways similar to the presence of a human companion (Schermerhorn et al. 2008). As such systems take on sophisticated human forms, it has also been argued that judgments of moral accountability increasingly come into play (Friedman and Millett 1995). These studies suggest that for robots to be successful assistants simple completion of tasks and efficacy is not enough. Rather, they need to exhibit naturalistic behaviors and appropriate emotional responses so that their human interactants will accept their environmental presence and be willing to engage.

Yet human communications are increasingly mediated by information technologies. Computer-mediated communication (CMC) has historically been framed as an impersonal phenomenon that encourages uncivil discourse such as flaming and trolling as well as group-based stereotyping due to deindividualization (Kiesler et al. 1984; Short et al. 1976; Spears and Lea 1992). Scholars have presented numerous theoretical frameworks to justify this stance. The Social Presence model argues that the fewer channels a medium has the lower the social presence afforded by the medium. Therefore, low-channel CMC makes it more difficult to build relationships than face-to-face communication since the former is perceived to be cold and impersonal while the latter is warm and sociable (Short et al. 1976). These conditions discourage communicative partners from seeing one another as potential friends because they lack the social cues and contexts that develop the conversational partner as a full complex person. In other words, "as bandwidth narrows, media allow less 'social presence'; communication is likely to be described as less friendly, emotional, or personal and more serious, businesslike, or task oriented" (Rice and Love 1987, p. 88). Another influential theory has been the Reduced Social Cues approach and Social Identity Model of Deindividuation Effects (SIDE), which argues that CMC's relative lack of social cues create an atmosphere where users may find it easier to express unpleasant sentiments and harsh decisions because they feel divorced from the human consequences of their actions. They do not have to witness the emotional fallout and distress caused by their words. These conditions also lead to heightened group identity and border patrolling. "Deindividuation theory proposes that behavior becomes socially deregulated under conditions of anonymity and group immersion, as a result of reduced self-awareness" (Spears et al. 2002, p. 94). SIDE theory postulates that when the individual's identity is not salient, the identity of the group becomes paramount and this in-group identity leads to stereotyping and ostracizing of out-group members. These theories all suggest that communications and interactions through technology would reduce investment of humanness on the interactants regardless of whether they were human or robot.

However, despite these limitations research shows that users find ways to increase CMC richness to achieve meaningful socially oriented communication. McCormick and McCormick (1992) show that communication partners can overcome the lack of face-to-face social cues through interpreting natural language speech patterns, questions and disclosures, or imbuing their messages with social meaning by using emoticons and punctuation. These methods augment the meaning of textual electronic messages in ways that convey some of the expressive contexts of verbal conversations (Walther and D'Addario 2001). Walther (1992, 1994) suggests that the social information-processing (SIP) model can explain these discrepancies. SIP argues that the impression development process takes longer in CMC, but given sufficient time the differences between CMC and face-to-face communications diminish since users have the capability of adapting the medium and find ways to overcome its limitations. This model recognizes that the nonverbal and lack of body language cues limit the scope of exchanges but by investing more time and messaging the relational effects into CMC can be brought to the same level as comparable face-to-face relationships. Communicative partners can read and recognize the humanity of the partner on the other side of the computer. This is important for understanding how we approach studying human interactions with robots because it is possible this process predisposes us to invest humanness in our CMC communicative partner even if that partner is not human. If the robotic communicator is able to properly mimic social cues, emotions, and appropriate responses this closeness could develop over an extended period.

Studies have also found that online communities can increase social ties and emotional support between users (Boase and Wellman 2006); the internet appears to be supplementing rather than supplanting prior human communications (Katz and Rice 2002), that online communities are useful for creating and maintaining weak tie networks (Ellison et al. 2007; Kavanaugh et al. 2005), and extending social interactions and supporting community building (Williams 2006). Even more, researchers have adopted the idea behind a sense of community—defined as a "feeling that members have of belonging, that members matter to one another and to the group, and a shared faith that members' needs will be met through their commitment to be together" (McMillan and Chavis 1986, p. 9)—to online environments, applying the term "sense of online community" (Katz and Aspden 1997; Quan-Haase et al. 2002; Wellman et al. 2003). Individuals who build and engage with communities online are also able to receive similar social and emotional gratification in offline settings as well. Thus, frequent online users are accustomed to communicating meaningfully with others whom they have never met in person, incorporating them into their community, and even developing emotional connections.

While traveling and elsewhere, we have become accustomed to technology mediating our communications and exchanges even in contexts that are in person and seemingly socially active. These different aspects—anthropomorphism of the robot, frequent meaningful communicative events over CMC with humans some of whom users have never met in person, and the increasing ways that nonhuman interactions mediate the everyday—may all be important for considering social outcomes of HRI and whether humans will accept robots as part of their social and environmental lives. These prior experiences inform the ways that users will conceptualize, respond to, and internalize interactions with robots.

3.2 Cultural and Demographic Differences in Robot Perception and Acceptance

Another important factor to consider may also be the cultural differences between groups and how this impacts acceptance of and interactions with robots (Halpern and Katz 2012). For example, in Japan, Shinto and Buddhist traditions both allow for conceptualizing robots as beings with spirit energy and the same vital forces that human beings have. There is also a long history in Japan of incorporating dolls into the social lives of individuals and even a Shinto shrine dedicated to providing proper funerals for dolls and housing their spirits. Japanese pop culture depictions of the future and alternate universes frequently include sentient robots that are integrated into the characters' social worlds and express a wide range of personalities and qualities many of which are positive (Shaw-Garlock 2009). Conversely, in America the most prevalent religious philosophies are Judeo-Christian, which does not allow space for inanimate objects to house souls. Shaw-Garlock also points out that when Western pop culture does depict sentient robots and artificial life it tends to focus heavily on the dangers of losing control over such creations and that this creates a deeply embedded cultural fear of machines. This fear, she argues, hinders acceptance of robots within Western cultures to the same extent as they are accepted in Japan (2009). However, it is worth pointing out that there are also dystopian visions of robotic futures and artificial life in Japanese anime such as the classic *Ghost in the Shell: Puppetmaster*. However, space considerations preclude further examination of this topic.

Differences within populations also play a role in how people respond to robots. For example, research has found that men tend to think of robots as more human-like than women do (Schermerhorn et al. 2008). Women instead tend to see robots as more machine-like and characterize them as less socially desirable. Nomura and colleagues conducted a separate study that yielded similar results in that women had more pronounced negative attitudes than men did toward situations involving interactions with robots (Nomura et al. 2009). These studies suggest that gender socialization may deeply impact interactions with robots, and therefore generalizations about cultural groups may be overly broad. Within any culture there are numerous subcultures and demographic differences that researchers need to pay closer attention to and directly address.

3.3 Research Studies

To better understand the ways that Americans currently interact with and respond to robots, and to test some of these suggestions, authors Katz and Halpern conducted a study (Halpern and Katz 2012; Halpern and Katz 2013; Katz and Halpern 2014). Though previous research has explored how people respond socially to robots' appearance and the effect of robots with humanoid aspects on attitudes toward robots, little has been done on individual level factors that moderate this relationship or how perceptions of human-like qualities in robots might relate to other background aspects of users such as use of certain information and communication technologies (ICT), gender roles, religious backgrounds, and previous experiences with robots. Americans do not belong to a homogenous culture nor do they all share the same experiences which means that it is important to understand the ways in which these individual level factors affect study results and what that might mean for the future of robots.

The first part of the study explored the relationship between recognition of human-likeness qualities in robots, social uses of ICT, and acceptance of robots based upon the following research question:

> For individuals, controlling for exposure to robot type, what is the relationship between activities that represent social uses of ICT and the willingness to accept robots as part of their social and physical environments?

This is based upon the idea that just as users adapt to a given electronic medium when they use it to communicate meaningfully with other users despite the lack of social cues, they will find similar ways to adapt to communications with robots (Walther 1997). The authors also predicted that users with a high sense of online community, high engagements with avatars, and a high level of perceived competence communicating with ICT would show a greater level of recognition of more human-like cues in robots. This will lead to relatively higher acceptance of robots as part of their social and physical environments.

In order to test this, a between-subjects empirical study was designed to identify human response to robots' appearances and their perceptions of robots. The participants were 789 undergraduate students (470 women, 283 men, 36 unidentified) enrolled in six communication courses at a large northeastern university. They ranged in age from 18 to 30 (M = 20.1, SD = 1.628). They were divided randomly into three groups and each group was exposed to an image of a different type of robot: Romeo, a French humanoid robot designed by Aldebaran to assist the elderly and disabled persons; the AIBO robotic dog designed by Sony, and an android with an extreme robotic appearance. The images of these robots were embedded within the questionnaire that participants completed online measuring their willingness to accept robots, the degree of human-like characteristics they perceived in them. The questionnaire also covered self-reports of their competence with ICT, engagement with avatars, and sense of online community. Demographic data was obtained to investigate topics of religion, gender, and age. The questionnaire utilized both dependent and independent scales. The dependent scales measured human-likeness and robotic social distance and the independent scales looked at competence with ICT, engagement with avatars in video games, and sense of online community. Religion, use of internet, use of *Second Life*, age, and gender were control variables.

Congruent with previous studies, participants who were exposed to humanoid robots recognized more human-likeness ($M = 3.25$, SD = 1.39), than in the android ($M = 2.93$, SD = 1.43) or the doggy robot conditions ($M = 2.73$, SD = 1.25). However, no significant differences were found between the android and doggy robot conditions. Nor was recognition of human-likeness alone in humanoid robots clear evidence those participants would be more likely to accept them. Results also show that participants with high sense of community ($F(1723) = 13.53$, $p < 0.001$) and individuals who engage more with avatars ($F(1731) = 48.5$, $p < 0.001$) both recognized significantly more human-likeness in robots. Additionally, engagement with avatars ($\beta = 0.138$, $p \leq 0.001$) as well as sense of online community ($\beta = 0.152$, $p \leq 0.001$) were both related positively to willingness to accept robots, even after controlling for demographic and robots' appearance. Interestingly, there was no significant effect for individuals who reported higher levels of competence communicating with ICT suggesting that just using ICT is not enough to change how individuals respond to or accept robots. Neither gender nor religion was associated with recognition or lack thereof of human qualities in robots but it did correspond strongly with acceptance. Running a SEM model confirmed that avatar engagement and a sense of online community were significant predictors of robotic social distance.

This study showed a few important things. First, exposure to humanoid designs increases recognition of human-likeness in robots but does not affect attitudes toward them. Participants' willingness to accept robots as part of their social worlds does not seem impacted by whether they recognized human-like qualities in those robots. The authors had hypothesized that following the same process of face-to-face human interactions, in which recognizing a person's identity and discovering similarities are relevant to developing social relationships (Kanda et al. 2004), recognition of human-likeness in robots would lead participants to accept them more. However, recognition of human-likeness could explain only 8 % of the variance, which means that more than 90 % of the willingness to accept robots depends on other variables not necessarily related to human-likeness. However, gender and religion were both strong predictors with men being more accepting and adherents of Judeo-Christian religions less likely to be accepting. This reinforces the idea that there are factors more important than exposure to visual stimulus of human-likeness when it comes to acceptance of robots in social and professional environments. Philosophical and cultural values held by respondents and cultural gender roles clearly impact comfort and acceptance levels concerning robots and their social routines. One limitation of this study was that all respondents were undergraduates at the same university, which limits the generalizability of the findings. Future research could explore the impact of religion, culture, and gender in-depth and tease out more nuanced factors that influence these issues from a wider demographic spectrum.

Second, this showed that while the use of ICT to communicate had little impact on willingness to accept robots, a sense of online community and avatar engagement did. Time spent using the internet as well as experience with *Second Life* (an online digital world where users interact with other human users through avatars) did not make a significant difference in acceptance. One possible explanation for this is that there is a difference between how frequently participants use this technology versus the ways in which they use that technology. Research suggests that informational and social uses of the Internet encourage community involvement and foster civic participation (Norris 2002; Shah et al. 2005), arguing that it is not time spent using a particular medium that makes a difference but rather how individuals use it (Norris 2002). In this study, the variables of using the Internet to get support from other peers, share knowledge, and meet others like them, and the feeling that the Internet made the respondent feel part of a larger community (all items represented in sense of online community) was positively associated with recognizing human-likeness. However, spending more time online did not necessarily indicate a higher recognition of human-likeness in robots. Likewise, individuals who indicated they engaged emotionally with avatars were more likely to recognize human-likeness, but not those who only interacted with avatars. This might explain the *Second Life* data since while there are opportunities to engage with other users' avatars it is minimal and the primary focus of the game is to create wealth and material goods for players' own avatars. In other words, emotional engagement with other avatars as full moral and emotional figures is not part of the average user experience in *Second Life.*

However, if the capacity to recognize human-likeness is indeed enhanced by more social use and engagement of ICT this study should have found a positive relationship between individuals who claim higher levels of competence communicating through ICT with others. Yet, the results did not show this either in recognition of human-likeness or attitude toward robots. One possible explanation is that increased exposure to technology and perhaps even robots in real life also increases respondents' awareness of their limitations. Tech savvy users have more realistic expectations for robots' capabilities, the ways that responses are part of their programs, and the kinds of tasks that robots excel at and ones they do not (Halpern and Katz 2012). Bartneck et al. (2005) reasoned along similar lines to explain why Japanese participants with a high degree of competence and experience using technology were more likely to indicate they felt uneasy about the idea of robots having emotions than less technologically savvy participants.

This second part of the study focused on exploring how attitudes about the suitability of robots for various occupations in society and how those attitudes might vary depending on robot appearance. The study also looked at individual level factors such as religion, gender, and experiences with several forms of technology. Two primary research questions framed this study:

> For individuals, controlling for robots' appearance, what is the relationship between gender, religiosity, and perceived competence with communication technologies, engagement with virtual reality environments, avatars, and attitudes toward robots?

> For individuals, controlling for types of robots, what is the relationship between attitudes toward robots and the occupations for which robots are believed to be qualified?

In order to answer these research questions, participants were asked to look at the robot (as was explained in the first part of the study) and then think about the needs that people in general have on a regular day. A list of 28 different occupations for robots was developed, and participants had to indicate how interested they would be in having the robot doing the different activities/occupations. The items were subjected to principal components analysis. The first component consisted of twelve items, which were occupations related to social and public assistant tasks (Cronbach $\alpha = .92$). We called this first factor Robots as Social Companions. Factor two, Robots as Surveillance beings, consisted of five occupations related to military and security tasks ($\alpha = .87$). The third factor, titled Personal Assistance, was composed of five occupations oriented to assisting subjects with different chores and household tasks. Then, subjects were asked to express how they would feel interacting with robots based on 25 items that show positive and negative aspects in this interaction. The items were subjected to principal components analysis and a three-factor solution was used. The first component Robot Liking (Cronbach $\alpha = .81$), consisted of eight items, each of which shows a preference for robots. Factor two, Cyber-Dystopian (Cronbach $\alpha = .71$), consisted of five items related to negative social consequences of the use of robots. The third factor titled Robotphobia (Cronbach $\alpha = .79$), was composed of seven items that mixed negative attitudes and emotions in the interaction with robots.

To test the relationship between the variables mentioned in our first research question of this second part of the study and attitudes toward robots, hierarchical multivariate ordinary-least squares (OLS) regressions were run to account for potential rival explanations and to assess the exact contribution of each block of predictors to the three factors constructed: Robot-Liking, Robotphobia, and Cyber-Dystopian. The total variance in robot-liking explained by the regression model was 27.1 %. The block of technological variables had a higher explanatory power compared to the demographic block, due to the strong relationship between avatar engagement and robot-liking. Recognition of human-likeness was also positively related to robot-liking ($\beta = 0.317$, $p < 0.001$), and even controlling by type of robots to which participants were exposed, subjects that recognized more human qualities in robots liked them much more than participants who recognized less human qualities. Recognition of human-likeness in robots was negatively correlated with robotphobia ($\beta = -0.148$, $p < 0.001$), suggesting individuals who do perceive human-like qualities in robots are less likely to hold negative attitudes and emotional responses in their interactions with them. Similar to the first study, Judeo-Christian religion was negatively related to Robot-Liking ($\beta = -0.176$, $p < 0.1$), but religiosity did not have a significant impact on it. However, religiosity was positively related to Robotphobia ($\beta = 0.061$, $p < 0.01$), which means that more religious individuals have a more fearful attitude toward robots. Gender also played a role with women negatively related to Robot-Liking ($\beta = -0.569$, $p < 0.001$) and positively associated with Robotphobia ($\beta = 0.555$, $p < 0.001$).

Interestingly, respondents with a high competence with ICT showed a positive relation to Cyber-Dystopia ($\beta = 0.104$, $p < 0.05$), indicating a generally negative attitude toward robots and supporting the explanation from study one. Individuals who have interacted with others in a MMORPG showed a marginally significant negative relationship with Cyber-Dystopian ($\beta = -0.314$, $p = 0.07$), which supports the idea that those who interact in virtual environments will show a more positive attitude toward robots. However, individuals highly engaged with avatars were positively related to Robot-Liking ($\beta = 0.152$, $p < 0.001$), but on the other hand, also positively related to Cyber-Dystopian ($\beta = 0.084$, $p < 0.05$), which means they have a positive attitude toward robots since they like them more but they also show higher concerns about the negative consequences in the use of robots by human beings.

To explore the second research question, hierarchical multivariate OLS regressions were run with the three factors constructed for occupations for robots as dependent variables, and the constructed factors for attitudes toward robots as independent variables. Not surprisingly, occupations related to social companion were positively associated with Robot-Liking, human-likeness, and avatar engagement. Occupations relating to surveillance were positively associated with Robot-Liking, negatively associated with robotphobia, and positively associated with Cyber-Dystopian.

This study reinforced many of the findings from the first study. Willingness to accept robots was not about the exposure to particular types of robots so much as recognition of human-like qualities of robots. This positive relationship can be explained through the concept of social presence, which is the sense that another intelligent being coexists and interacts with the user in the same environment (Biocca 1997). Research has shown that users' attitudes, evaluations, and social responses toward robots are mediated by their feelings of social presence during their interaction with robots (Lee et al. 2006). Respondents who are highly familiar with ICT and spend a lot of time engaging with these technologies were also more likely to fall into the Cyber-Dystopian camp while users who spent more time engaging with avatars were more likely to fall into the Robot-Liking camp. Adherents of Judeo-Christian traditions were also negatively correlated with Robot-Liking. And gender enculturation seems to play a significant role in acceptance.

The study also tested attitudes toward robots fulfilling certain occupations in society. It found that individuals who like robots prefer them to do activities related to social companionship and surveillance functions. But they interestingly did not have a significant preference for them as personal assistants. These findings support that of Goetz et al. (2003) which found that people systematically preferred robots for jobs when the robot's human-likeness matched the sociability required for those jobs. They also found that individuals felt robotic assistants needed to exhibit naturalistic behaviors and appropriate emotions with little to no learning effort on the user's end. Neither of these characteristics and features was embedded in the robots presented in our study, so it is possible to argue that more realistically designed robots would have elicited from participants stronger preferences to use them as personal assistants. Respondents with Robotphobia not surprisingly showed lower preferences for robots fulfilling any of the occupations. But those with a Cyber-Dystopian attitude did prefer robots for surveillance.

This may be due to a few reasons. Perhaps they perceive this dystopian threat as coming from without and robot surveillance would help protect them. Or they may believe that when retained in a security capacity the robots are more likely to remain under human control. Conversely, their view of a dystopian future may map onto the numerous science fiction presentations of a world where every aspect of our lives are recorded and robot surveillance is simply a potential they accept.

3.4 Conclusion

Earlier, we identified three factors that were important for considering social outcomes of HRI—anthropomorphism of the robot, frequent meaningful communicative events over CMC with humans some of whom users have never met in person, and the increasing ways that nonhuman interactions mediate the everyday—along with a few cultural aspects such as religion, gender, and social attitudes. The two studies touched on some of these points and it is worth considering how they might inform future studies. Both studies found that anthropomorphism of robots did effect how people responded to them, with more human-looking robots being attributed as having more human-likeness. However, they also revealed that willingness to accept robots was more complex than simply how those robots appeared. Different experiential and sociocultural factors predisposed people to accepting them. Rather surprisingly, higher exposure to ICT can elicit a dystopian attitude toward robots. Perhaps the mediation of nonhuman actors in our everyday lives can create a negative effect. But using certain kinds of avatars increased acceptance suggesting that meaningful communicative events over CMC might positively impact acceptance. Gender also played a role in that women were less likely to accept robots in their social and work environments. Lastly, religion and religiosity proved to be a factor supporting the idea that Judeo-Christian traditions do not allow space for robots as social beings.

This research also highlights areas that need more attention in academic studies of robot human interactions. Clearly cultural factors are very important but our study was only able to examine undergraduates in a particular American context. Future research should explore how other cultural groups conceive of and respond to robots, which would serve two important purposes. First, as industry increasingly seeks to incorporate robots into our everyday lives and governments seek to develop policies regarding this it is vital that we understand not all societies will respond in the same ways. In the European Union, for example, there are a multitude of unique and thriving cultures that each approaches the world through different perspectives. Therefore a "copy-paste" tactic may not be effective. Instead, research on local concepts and attitudes should inform the design, programmed responses, and interactive styles of the robots as well as the accompanying policies regarding robots will need to adjust to these local realities. Second, cross-cultural studies provide a way for us to see which aspects might *not* be cultural. In other words, if there are constants across a wide variety of cultures we may be able to isolate factors that are common to all humans and would allow us to hypothesize about biological and evolutionary origins of attitudes and responses. However, in order to confidently make any such claim numerous rigorous and wide ranging cross-cultural studies would need to be conducted.

Likewise, the role of gender and religion merit further exploration. Gender socialization varies widely across cultures so that the divide between young women and men in America may not necessarily characterize the situation in other contexts. For example, Sweden's commitment to gender equality and reducing the gender gap economically and educationally may also greatly reduce the gender gap in attitudes toward HRI (and research in that country would possibly shed light as well on the questions raised in the preceding paragraph). Regarding religion, previous research including our own only looked at Judeo-Christian, Shinto, and Buddhist traditions, by-passing other major religions such as Islam and Hinduism. It is also important to note that within these larger categories are numerous denominations and sects, which may approach these topics differently. For example, scholarship suggests that the soul status of nonhumans has long been a debate within Christianity and that even if religious institutions issue edicts regarding the issue practitioners may view the topic differently (Preece and Fraser 2000). A deeper exploration of the role of religion and religiosity in robot acceptance and attitudes would therefore be worthwhile.

Another useful line of inquiry is the appearance of the robot and the type of task it is programmed to complete. There are cultural concepts of how particular types of people and bodies are better or worse suited for tasks. For example, a robot tasked with lifting heavy objects might be expected to appear squatter. A gracile robot might strike observers as inappropriate for the task even if it was perfectly able to complete the required undertakings. (Certainly there is a substantial body of literature demonstrating the importance of appearance when humans consider other humans for a variety of tasks.). There may also be physical expectations that differ between robots who engage in a more social manner and those engaging in undesirable tasks. The other side of this aspect is that robot creators should be sensitive to these cultural concepts as it might relate to delicate matters such as race, class, and caste. Biases and histories of inequality of particular demographic groups often include stereotypes about appearance. If a robot is programmed to perform a particular task that is associated with a marginalized group, then a culturally sensitive approach should be employed when considering physical design as well as other dimensions such as interaction patterns, voice, and color.

Forms of technologically mediated interactions also deserve further investigation. Our research suggests that previous use of avatars might impact how willing people are to accept robots but we did not separate out particular types of avatars aside from *Second Life*. Video games routinely provide ways for players to design avatars to their liking and then interact with programmed characters and the avatars of other users. Yet, not all video games place heavy emphasis on emotional and moral character development or social interactions with avatars. First person shooting games, for example, often have online components where users play with and against avatars of other human users. These formats have historically not provided interaction contexts that foster development of meaningful conversations and emotional attachment. Teams can be randomly assigned, the game can be heavily competitive, and users can always respawn to play again. But this has changed in recent years, in part because of growing bandwidth and availability of hardware, a trend that further reinforces the idea that people like social situations and interacts with humans and human-like entities. Other types of virtual games, often story driven, include programmed characters that can interact with the human users in a variety of ways and elicit emotional attachments despite users' knowledge that they are fictional nonhumans. The degree that a game encourages emotional investment in characters and avatars could be an important factor for how users later think about engaging with nonhuman actors like robots.

Lastly, though this research was not specifically examining the uncanny valley it does suggest that its impact may not remain as powerful as it is currently. This phenomenon might not be due to a fault in the concept of the uncanny valley so much as a testament to how plastic humans can be when adjusting to new realities. In 1897, when Edison showed a grainy, silent, black, and white film of an on-rushing train, called the *Black Diamond Express,* it elicited such fear from audience members that they bolted from the auditorium. The so-called "train effect" where audiences panicked when presented with film of approaching vehicles occurred repeatedly during that era in Europe, Shanghai, and America. Viewers responded by flinching, wincing, yelping, and even running away (Bottomore 1999). This was not an irrational response but merely the equivalent of the uncanny valley of their time. Actually, people can overcome these discomforts rather quickly and adjust to the new normal. We as a species can adapt quite rapidly to changing cultural contexts and visual media so that even if robots are initially too like us for comfort, most of the population may be able to normalize and adjust rather swiftly. Therefore, initial experimental responses regarding both the uncanny valley and anthropomorphism in general might not indicate future attitudes. A longitudinal study could explore whether cultural groups can adjust to particular appearances and interaction formats successfully and whether they would grow to accept robots in their social and work environments.

References

Bartneck C, Nomura T, Kanda T, Suzuki T, Kato K (2005) Cultural differences in attitudes toward robots. In: Proceedings of symposium on robot companions: hard problems and open challenges in robot-human interaction, SSAISB 2005, pp 1–4

Biocca F (1997) The cyborg's dilemma: progressive embodiment in virtual environments. J Comput-Mediat Commun 3(2). http://jcmc.indiana.edu/vol3/issue2/biocca2.html. Accessed 10 Jan 2012

Boase J, Wellman B (2006) Personal relationships: on and off the internet. In: Perlman D, Vangelisti AL (eds) Handbook of personal relations. Blackwell, Oxford, pp 709–723

Bottomore S (1999) The panicking audience? Early cinema and the 'train effect'. Hist J Film, Radio Telev 19(2):177–216

Duffy B (2003) Anthropomorphism and the social robot. Robot Auton Syst 42:177–190

Ellison NB, Steinfield C, Lampe C (2007) The benefits of facebook "friends:" social capital and college students use of online social network sites. J Comput-Mediat Commun 12:1143–1168

Fong TW, Nourbakhsh I, Dautenhahn K (2002) A survey of socially interactive robots: concepts, design, and applications. Robot Auton Syst 42:142–166

Fortunati L, Katz JE, Riccini R (eds) (2003) Mediating the human body: technology, communication and fashion. Erlbaum and Associates, Mahwah

Friedman B, Kahn PH, Hagman J (2003) Hardware companions? What online AIBO discussion forums reveal about the human-robotic relationship. In: Proceedings of CHI '03: SIGCHI conference on human factors in computing systems

Friedman B, Millett L (1995) It's the computer's fault—reasoning about computers as moral agents. In: Proceedings of the conference on human factors in computing systems (CHI 95), ACM Press, Denver, CO, pp 226–227

Goeatz J, Kiesler S, Powers A (2003) Matching robot appearance and behavior to tasks to improve human-robot cooperation. In: Proceedings of robot and human interactive communication, 2003, Roman

Halpern D, Katz JE (2013) Close but not stuck: understanding social distance in human-robot interaction through a computer mediation approach. http://www.fus.edu/intervalla/images/pdf/3_halpern_katz.pdf

Halpern D, Katz JE (2012) Unveiling robotophobia and cyber-dystopianism: the role of gender, technology and religion on attitudes towards robots. In: Proceedings of HRI '12: 7th international conference on human-robot interaction

Kanda T, Hirano T, Eaton D, Ishiguro H (2004) Interactive robots as social partners and peer tutors for children: a field trial. Hum-Comput Interact 19(1): 61–84

Katz JE (ed) (2003) Machines that become us: the social context of personal communication technology. Transaction Publishers, Piscataway

Katz JE, Aspden P (1997) A nation of strangers? Friendship patterns and community involvement of Internet users. Commun ACM 40(12):81–86

Katz JE, Halpern D (2014) Attitudes towards robots suitability for various jobs as affected robot appearance. Behav Inf Technol 33(9):941–953. doi:10.1080/0144929X.2013.783115

Katz JE, Rice R (2002) Social consequences of Internet use: access, involvement and interaction. MIT Press, Cambridge

Kavanaugh A, Reese D, Carroll J, Rosson M (2005) Weak ties in networked communities. Inf Soc 21:119–131

Kiesler S, Siegel J, McGuire TW (1984) Social psychological aspects of computer-mediated communication. Am Psychol 39:1123–1134

Lee KM et al (2006) Can robots manifest personality? An empirical test of personality recognition, social responses, and social presence in human–robot interaction. J Commun 56:754–772

McCormick NB, McCormick JW (1992) Computer friends and foes: content of undergraduates' electronic mail. Comput Hum Behav 8:379–405

McMillan DW, Chavis DM (1986) Sense of community: a definition and theory. J Commun Psychol 14(1):6–23

Nomura T, Kanda T, Suzuki T, Yamada S, Kato K (2009) Influences of concerns toward emotional interaction into social acceptability of robots. In: Proceedings of HRI '09: 4th ACM/IEEE international conference on human robot interaction, La Jolla, CA, pp 231–233

Norris P (2002) The bridging and bonding role of online communities. Harv Int J Press/Pol 7(3):3–13

Preece R, Fraser D (2000) The status of animals in biblical and Christian thought: a study in colliding values. Soc Anim 8(3):245–264

Quan-Haase A, Wellman B, Witte J, Hampton K (2002) Capitalizing on the internet: network capital, participatory capital, and sense of community. The Internet in Everyday Life. Oxford, Blackwell, 291–324

Rice RE, Love G (1987) Electronic emotion: socioemotional content in a computer-mediated network. Commun Res 14:85–108

Schermerhorn P, Scheutz M, Crowell CR (2008) Robot social presence and gender: do females view robots differently than males? In: Proceedings of the 3rd ACM/IEEE international conference on human robot interaction (HRI 2008), Amsterdam, NL, pp 263–270

Shaw-Garlock G (2009) Looking forward to sociable robots. Int J Soc Robot 1:249–260

Shah DV, Cho J, Eveland WP, Kwak N (2005) Information and expression in a digital age: modeling internet effects on civic participation. Commun Res 32:531–565

Short J, Williams E, Christie B (1976) The social psychology of telecommunications. Wiley, Hoboken

Spears R, Lea M (1992) Social influence and the influence of the "social" in computer mediated communication. In: Lea M (ed) Contexts of computer-mediated communication. Harvester-Wheatsheaf, London, pp 30–65

Spears R, Postmes T, Lea M, Wolbert A (2002) When are net effects gross products? The power of influence and the influence of power in computer-mediated communication. J Soc Issues 58(1):91–108

Walther JB (1992) Interpersonal effects in computer-mediated interaction: a relational perspective. Commun Res 19:52–90

Walther JB (1994) Anticipated ongoing interaction versus channel effects on relational communication in computer-mediated interaction. Hum Commun Res 20:473–501

Walther JB (1997) Group and interpersonal effects in international computer-mediated collaboration. Hum Commun Res 23:342–369

Walther JB, D'Addario KP (2001) The impacts of emoticons on message interpretation in computer-mediated communication. Soc Sci Comput Rev 19:324–347

Wellman B, Quan-Haase A, Boase J, Chen W, Hampton K, Isla de Diaz I (2003) The social affordances of the Internet for networked individualism. J Comput-Mediat Comm 8(3). http://jcmc.indiana.edu/vol8/issue3/wellman.html

Williams D (2006) On and off the 'net: scales for social capital in an online era. J Comput-Mediat Comm 11:2

Zhao S (2006) Humanoid social robots as a medium of communication. New Media Soc 8:401–411

Chapter 4
Perception, Acceptance, and the Social Construction of Robots—Exploratory Studies

Joachim R. Höflich and Afifa El Bayed

In order to investigate social robots' images and approval and/or rejection, we conducted two pre-studies and two studies combining qualitative and quantitative methods. In the first pre-study, we attempted to capture the images of robots in our minds as indicators of their social representation. The second pre-study tried to contrast the acceptance of robots in general among non-experienced users with their acceptance of specific robotic instances. In the first study, we aimed to quantitatively shed light on the relationship between the acceptance of robots. The users' demographics and their earlier robotic experience on one hand, and the robots' functionalities, interactional capacities, and physical appearance on the other. Finally, in the last study, we aimed to track the shift from a dyadic to a triadic relationship with a robot by observing the interaction between pairs of participants and a robot in an experimental setting. Our four studies provide support for previous academic findings, but also question others opening up the floor for novel research horizons.

4.1 Introduction: Robots as Unknown Entities and as Pictures in Our Minds

For many years, the computer has been considered the basal medium of modern societies. Nowadays, as a recent Pew Research Center (2014) study underlines, a change is in progress that replaces computers with robots. Indeed, the majority

J.R. Höflich · A. El Bayed (✉)
University of Erfurt, Erfurt, Germany
e-mail: a.el_bayed@uni-erfurt.de

J.R. Höflich
e-mail: j.hoeflich@uni-erfurt.de

J. Vincent et al. (eds.), *Social Robots from a Human Perspective*,
DOI 10.1007/978-3-319-15672-9_4

of the respondents declared that artificial intelligence in general, and particularly robots, will determine a large part of life in the year 2025 "with huge implications for a range of industries such as health care, transport and logistics, customer service, and home maintenance". Although everyone knows 'robot' as a term, only few have concrete experiences with one. Based on data gathered from citizens of the European Union (European Commission 2012, p. 14), 87 % of Europeans declared that they have never used a robot in their lives, neither at work nor at home (including robotic vacuum cleaners). In this sense, the experience with robots remains for many as a myth presented by science fiction literature and films, instead of being a self-experienced encounter. Hence, fears of witnessing such machines becoming virulent exist and persist (cf. Neilsen 2011).

Fear is a barrier regarding the acceptance of robots and measuring the mental states of users and non-users of robots drives research, especially that which explores attitudes toward robots using "Negative Attitudes toward Robots Scale" (NARS) (Nomura et al. 2006a, b, c; Tsui et al. 2011; Syrdal et al. 2009). Attitudes are a kind of substitute for behavior; a pre-behavior associated with the idea to predict (and also change) behavior. Based on a dyadic perspective of a human–robot interaction and an atomistic perspective, as Moscovici (2002, p. 234) indicates, "social psychologists studying attitudes are not really interested in people's knowledge in their symbolic world." Following Moscovici, to acquire an attitude towards an object means to have a representation, which is part of a culture of folk knowledge and of cognition. Social representations do not substitute or oppose attitudes, but rather they are part of them. These social representations are 'social' as far as they are based on communication. In this sense, the perception and acceptance of robots are based on attitudes that are part of representations that are constituted by communication. In other words, robots as well as media are social constructions in a process of social incorporation—a process, that indicates that the robots are not social in nature, but rather gain their sociability thanks to processes of social appropriation. This makes robots a 'robotic other' that is part of a triangular conception of representation, where Person A and Person B relate to the robot and include it or exclude it from their interaction.

This chapter presents exploratory studies reflecting the idea of the social representation of robots, where robots are seen as media, socially constructed by communication and socially embedded. The paper ends with some conclusions regarding further fields of research.

4.2 Pre-study 1: The Perception of Robots; Their Pictures in Our Minds

The picture of robots many have is limited to their imagination and in this pre-study we aimed to discover what these social representations of robots look like, and whether these perceptions comply with the marketed ones. Since April 2011 we have been collecting drawings of robots in order to capture our participants' mental images and to reveal their perceptions of them. The (ongoing) sample includes

Fig. 4.1 Imaginations of robots

students of both genders, enrolled in different humanities and social sciences programs at the University of Erfurt, with an age varying between 18 and 25 years. This exploratory trial is based on the idea that "Representation = image/meaning, that it equates every image to an idea and every idea to an image" (Moscovici 2002, p. 31). This is associated with the premise that the images of robots in our minds can be to some extent reflected and made visible in drawings of robots. Accordingly, the students participating in this study were given the instruction, without additional guidance, to draw a robot.

Figure 4.1 shows a few of the many drawings by our participants which demonstrated the following characteristics:

- Robots were frequently shown as they appear in book illustrations, following the typical cube-model pattern. This actually connects robots in our minds to the antiquated image of robots which is not found or followed in the contemporary robots of today. In the same spirit, the drawn robots do not differ from those sketched by children over 35 years ago (see Reichardt 1978, p. 100).
- Robots often appear to rather resemble men (masculine robot model), while…
- … female robots appear gender-stereotyped.
- Male robots are often associated with dominance attributes, if not with some potential for violence.
- Interestingly, a lot (although not the majority) sketched an on–off button for the robot. This highlights the wish to keep control over the robot when needed, so that their movement and activities can be limited.

Taking into consideration these findings, it is clear that our study was able to show that persons with limited or no knowledge about robots have a perception about them that is based on their imagination. It also underlined that the imagined robots remain caricatured in nature and rather science-fiction-movie-character-like, and do not resemble those developed in laboratories, or that are marketed (e.g., Aibo as a zoomorphic robot, Kismet as a functionally designed robot). The participants' perceptions about robots remain mainly fictional in nature and incompatible with reality. All this in an age when robotics is an increasingly flourishing discipline, and when robots are increasingly encouraged to become socially incorporated.

4.3 Pre-study 2: The Acceptance of Robots; Those We Know and Those We Imagine

In this second pre-study, we decided to go a step further with the aim of unearthing the general acceptance of robots as imagined and perceived before exposure to real instances of existing robots. Hence, in September 2014 we ran a further qualitative study, which took the form of a problem-centered qualitative interview. The 26 persons sampled had either no prior experience with robots or a limited one. They were students of both genders (n (Male) = 13; n (Female) = 13), belonging to different humanities and social sciences undergraduate and graduate programs at the University of Erfurt. Some also occasionally work part-time, and others attend training programs.

In this pre-study, the participants were asked about their acceptance of robots in general as well as about their incorporation in society, before being asked to answer open-ended questions about their acceptance of robots holding different social roles in various social contexts (e.g., housekeepers, nursing assistants in hospitals, mine extractors in the field of war). After being asked about robots in general, and about robots performing specific tasks, they were then exposed to real instances of robots and were asked about their acceptance.

The results of this second pre-study revealed interestingly that our respondents—even with their lack of experience with robots—were ready to accept social robots in general. Still, they accepted some more than others in specific environments. Our participants were open to the introduction of robots in contexts where their inclusion brings a benefit—an added value. They even welcomed their replacement of humans in some specific tasks and duties (e.g., repetitive tasks to avoid human boredom, missions in dangerous fields to guarantee human safety). On the other hand, robots were rejected in cases where human beings proved to be more competent. In other words, it is generally believed that humans are more qualified than robots in certain contexts, for example medical care, and so should remain. When it came to robotic performances involving interaction with human counterparts, these were generally underestimated and looked down upon such as in this example for a male respondent, "*I find that a robot should not be used in areas where it depends on the interpersonal interaction.*" This statement, being declared by a teacher, goes hand in hand with previous findings indicating the fear of robots taking people's jobs.

The second part of the results, which concerns the responses of the participants regarding the specific robots to which they were exposed, noted the increase in the openness of the participants; in fact they were more accepting of the concrete instances of the robots they were confronted with. For instance, they were more tolerant to RI-MAN (a robot designed for medical assistance) than they were to medical care robots in general. However, we noted that the level of acceptance dropped as the participants were evaluating robots that were very physically similar to human beings.

The findings of pre-studies 1 and 2 demonstrated that our participants, who lacked knowledge and earlier experience with robots relying mainly on their imagination, generally accepted robots including into their social life. But when given the chance to think about the main aspects the robots might deal with and the tasks they should fulfill, people started giving privilege to humans and underestimated potential robotic performance in tasks that are or they felt should be kept human in nature. This highlights the mental line that persons draw between what is human (e.g., medical care, teaching) and what should become robotic (e.g., mine searching, mechanical assembly). This simply underlines the fear of some kind of robotic control (a classical science fiction scenario) that might cause human uselessness or simply increase unemployment (European Commission 2012, p. 149).

However, when confronted with real instances of robots, our participants showed more flexibility and acceptance of robots, even in fields where they believed that robots would be rather useless (e.g., medical care). This sheds light on the shift in perception that acceptance of robots undergoes when knowledge about them increases. It is important to note that with respect to the physical appearance of the selected robots, our participants' acceptance of robots increased as their appearance became more human-like, before it radically dropped as their resemblance became very human-like. This finding supports Mori's uncanny valley hypothesis (1970, 2012), which attests that the increase in physical resemblance of robots to humans engenders an increase in their acceptance until they become so human-like it causes a drop in acceptance or even rejection.

4.4 Study 1: The Perception and Acceptance of Robots—A Survey

In the light of the results of our two pre-studies, we conceived a quantitative study that aimed to capture the perception and acceptance of robots by persons with limited or no knowledge about them, and by those who have earlier experience with the robots, to find out the difference between these, if any. Hence, we sampled between 2011 and 2014 a number of persons ($n = 130$), of whom there were slightly fewer men than women, comprising: high school students (14.6 %), college students (35.4 %), employees (31.5 %), independently working (6.9 %), retired (7.7 %), and unemployed persons (3.9 %).

The survey was conducted online and paper-based copies were also used by some for convenience. The survey questionnaire was designed to address the

Fig. 4.2 Robots used in our questionnaire

following research questions based on the results of our pre-studies and supported by earlier research in the field (Reeves and Nass 1996).

RQ1 What is the relationship between the previous contact with robots and their acceptance?
RQ2 What is the relationship between the complexity of a robot's functionalities and its acceptance?
RQ3a What is the relationship between the interactional similarity of a robot with human beings and its acceptance?
RQ3b What is the relationship between the physical similarity of a robot with human beings and its acceptance?
RQ4a What is the relationship between a person's age and the acceptance of robots?
RQ4b What is the relationship between a person's gender and the acceptance of robots?
RQ4c What is the relationship between a person's educational level and the acceptance of robots?

We designed our questionnaire in five blocks. The first block was self-redacted and asked the respondents about their earlier experience with robots as well as the field of operation of the latter (e.g., industry, robot toys, home assistants, museum tour guides). Then, they were asked to evaluate their acceptance of robots using a five point Likert scale according to the robots' trustworthiness, helpfulness, entertaining, simplicity, and pleasance. This first cluster of questions enabled addressing our first research question RQ1.

Question blocks two, three, and four were similar but referred to different robots. In each block, the respondents were shown a picture of a specific robot (see robots a, b, and c in Fig. 4.2), and were asked to estimate their acceptance

Table 4.1 Independent samples *t*-test comparisons of means for items evaluating HRI by persons with or without experience with robots

		Trustworthiness	Helpfulness	Pleasance	Entertainment
	n	*M* (SD)	*M* (SD)	*M* (SD)	*M* (SD)
People with earlier experience with robots	66	3.98 (0.75)	3.26 (0.80)	4.45 (0.50)	3.32 (0.84)
People without earlier experience with robots	64	2.08 (0.80)	2.95 (0.68)	2.38 (0.52)	2.95 (0.68)
t-test	–	13.96***	2.35*	23.22***	2.71**

***$p < 0.001$; **$p < 0.01$; *$p < 0.05$

of the robot in different aspects and contexts of social life (e.g. medical clinic, garden, neighbors house, bedroom) using a five-point Likert scale.

The second block of questions addressed Robot 'a' as a representative example of mechanical robots. The third addressed Robot 'b' as an instance of social-interfaced robots and the fourth asked questions about a humanoid robot Robot 'c'. As one considers the robots 'a', 'b', and 'c' respectively, one notices that the complexity of their functionalities and their interactional and physical similarity with humans increases. These four blocks of questions thus address our research questions RQ2, RQ3a, and RQ3b.

Finally, the fifth block asked the participants about their demographics such as their gender, age, and educational level. Combined with the indicators on the acceptance of robots presented above, they addressed research questions RQ4a, RQ4b, and RQ4c.

Coming to the results of our study, we succeeded to sample persons among which around 50 % had earlier experience with robots in the form of robot-toys (48 persons), industrial robots (41), robots exposed in museums (34), in the workplace (23), in commerce (18), and home assistants (22). This quota particularly addressed RQ1, which explored the relationship between previous contact with robots and their acceptance. As our results in Table 4.1 show persons with earlier robot experience positively described them as being trustworthy, helpful, pleasant, and entertaining when compared with persons with no previous robotic experience.

In research question RQ2, we asked about the relationship between the complexity of the robots functionalities and their acceptance. And in RQ3a and RQ3b, respectively, we aimed to examine the relationship between the interactional and physical similarity of robots with humans and their acceptance. The results showed that the more complex the functionalities of a robot were and the more interaction and physically human-like it was, the less accepted it was (see Table 4.2). In our computations, the acceptance of a robot was considered to be the sum of its acceptance in

Table 4.2 Comparisons of the means for the acceptance of the robots a, b, and c

		Acceptance
	n	*M* (SD)
Mechanical robot 'a'	129	4.05 (0.82)
Social-interfaced robot 'b'	128	2.59 (0.98)
Humanoid robot 'c'	129	1.59 (0.72)

$F(2, 383) = 292.70, p = 0.000$

the different social life aspects as evaluated by our participants divided by the total number of items. After attaining a significant between-group means comparison of the acceptance of the robots ($p < 0.001$), a Tukey post-hoc test was run to demonstrate the significance in the decline of acceptance as the robot's similarity with humans in terms of complexity, interaction, and physical appearance increases (Robot a's acceptance * ($p < 0.05$) > Robot b's acceptance * ($p < 0.05$) > Robot c's acceptance).

Research questions RQ4a, RQ4b, and RQ4c tried to capture the relationship between the demographics of the participants in terms of gender, age, as well as educational level and their acceptance of robots. Our results showed that in general (but not significantly) men accepted robots more than women; younger people demonstrated more acceptance toward robots than older ones, and more educated persons were more robot accepting than less educated ones.

In sum, our results compare favorably with conclusions previously reached by several Human-Robot Interaction scholars. Earlier research in the field supports the idea that user-friendliness (Isbister and Nass 2000), simplicity (Norman 1988) and consistency (Winograd 1996) of robotic platforms increases their acceptance (Isbister and Nass 2000), as opposed to their complexity which causes its decrease. Similarly, earlier experience with robots is believed to increase their acceptance (Reeves and Nass 1996), a finding we reached in the present study, as well as in our second pre-study.

On the other hand, some of our results do not necessarily align with former findings in the field. Although it is believed that human beings prefer, and even continuously seek, interaction with similarly communicating counterparts (Reeves and Nass 1996), our study found that the more the robotic counterpart was interactionally similar, the less it was accepted. Coming to the physical aspect of the robot and its impact on the interaction, the mostly supported hypothesis in the field remains the uncanny valley by Mori (1970, 2012). In our studies the acceptance of the robots decreased whenever their physical resemblance with humans increased, showing a categorical disfavoring of anthropomorphism, of which many scholars including Duffy (2003), Ray and Siegwart (2008), and Turkle (2011) argue in favor. Indeed, it seems that our participants tended to reject robots whenever they increasingly resembled them, underlining hence a form of fear of the robotic other. This fear, that is described by Mori (1970, 2012) and which he believed should appear only in further stages of human-likeness, showed up in our case at an earlier stage, emphasizing consequently the difference in the way humans perceive robots as the intensity of experience with them varies across cultures (in the current case, German participants versus Japanese robots). This might also be the reason why the demographics did not appear to be a factor in the acceptance of robots in any remarkable way, since our sample includes persons (even those with earlier experiences with robots) who equally never experienced living in a setting where robots were socially imbedded, so that their gender, generational or even educational levels come into play. In fact, Nass and Moon (2000) remind that situational factors also may influences acceptance. This encouraged us to pursue a second study in a different context that placed two persons with a social robot in an experimental setting.

4.5 Study Two: The Social Construction of Robotic Others

When people engage in a relationship with a robot, they engage in a dyadic relationship on an 'as-if' basis of imagination; as if the robot was a social being. The starting idea is that robots become social robots when they are socially embedded and incorporated. In this case, the robot should not only be in a situation whereby it stands face to face with a person (and being hence regarded as a social being), but should rather be considered as a building block of a social fabric; a social context. In this case, the robot cannot be perceived anymore as an object, but rather as an interlocutor, and not as a third object, but as a third person who can at every moment integrate the communication structure (Höflich 2013). One can also speak about a 'robotic third', or in the words of Kahn et al. (2004, p. 546) about a 'robotic other'. With this term, Kahn et al. (2004, p. 546) try to go beyond the ontological definition of a social robot to point more to how people can designate a robot from a social perspective. Following this idea, one may also refer to the German sociologist Georg Simmel (1995) and his conception of the third member. The third indicates an extension of a dyadic relation towards the social/sociological dimension and the emergence of a structure. Looking at triadic relationships associated with a third is the possibility of incorporation as well as of exclusion. In this sense, a robot could be an integral part of the triad or it could be excluded, to be considered as an external artifact.

The others in interaction must be classified—and consequently the way in which they transform an object to a constructed quasi-subject must be traced. The starting point is the imagination of objects as social objects. This has to do with the core idea of symbolic interactionism. With this respect, Herbert Blumer (1986, p. 11) points out that objects do not carry meaning in themselves, they rather acquire it in a (continuous) process of mutual displaying. Hence, they become necessarily meaningful objects.

In this process of social interaction, social representation emerges. This underlines that humans are not "lonely cognizers" (Wagner and Hayes 2005, p. 119), rather the relation of a person to an object (in this case; to a robot) is defined and mediated by his or her relations to others. However, this object has a particularity, which is that it can be (more or less autonomously) interactive. It reacts to our actions and we react to the one it emits. Seen this way, robots can be considered as 'interactive media' (Zhao 2006), with which we interact and that constitute communicative bridges, when it reads email, or as in the case of Paro (a seal robot used for elderly care) when it stimulates communication among patients suffering from dementia.

This second case study takes rather the form of an experimental arrangement embracing methods such as think aloud, observation and a survey. Although the current results are at a preliminary status, they already introduce important findings that open up a fundamental discussion regarding the current perception of robots and its potential changes. The current observational study permitted the occurrence of an experimental triadic arrangement, with two individuals and a robot, that focused on and looked into the details of the communicative embedding (and by the same occasion the inclusion) of a robot. Due to limited financial support, we opted for an affordable robot, which still had the necessary features for our experiment. This toy robot is Robosapien (toy of the year 2004). Produced by WowWee and developed by the NASA-physicist Mark Tilden, the robot is promoted as follows:

> Robosapien is a sophisticated fusion of technology and personality. Loaded with attitude and intelligence, ..., Robosapien is more than a mechanical companion—he's a multi-functional, thinking, feeling robot with attitude! (http://wowwee.com).

The participants received written instructions beforehand and were asked to teach the robot how to fulfill tasks such as walking without encountering obstacles and raising an object with its hand (modeled to enable gripping).

In each experimental event, two persons were involved, totalling eight pairs (4 male, 4 female), all students of the University of Erfurt. Each couple was instructed to articulate their thoughts out loud during the experiment, in order to capture their thoughts and perceptions. The discussions involving our participants were recorded, transcribed, and content analyzed and a semi-structured interview was conducted and recorded with participants. Figure 4.3 below gives a general impression of the experiment.

The idea behind the experiment was to grasp the 'social' in the robot—as a social construct—based on the descriptions, which emerged during the discussions and that designated the robot as a member of the social setting. In this case, the robot is considered as a part in a triad, following the taxonomy of human-robot interaction as proposed by Yanco and Drury (2002). They examined relationships between one or two persons and/or robots demonstrating how the relationships vary as the persons intervene as single instances in the interaction with the one or many robots, or when they intervene as a group with the one or many robots. These two constellations proved to be relevant to our experiment, especially as the conversation of our paired subjects continuously shifted from a direct interaction

Fig. 4.3 Instance of the experiment with Robosapien

with the robot (inclusion) to an interpersonal communication showing distance to the robot (exclusion).

Robosapien was not referred to as research object, not a robot, by our laboratory supervisors. This was to avoid any influence or bias of our respondents' perceptions, but as soon as the pre-briefing started the assignment of specific names occurred. The robot was referred to with respect to characters in movies like C-3PO (known from the Star Wars movie). Nicknames such as "Robbi-Bobbi", "Robbi", "Bunny" or even "Schatzi" (German word equivalent to 'my little treasure') were used; also, neutral references like "object". Using names, nicknames or natural references is a proof of the need to provide the robot—the initially presented plain object—with a name. By naming the object, the foundation for a personal reference is set. Giving a name indicates an anchor for social representations, "to set them in a familiar context" (Moscovici 2002).

However, one should not forget that Robosapien was also perceived as a toy, because it is one and reactions to it were provided using human attributions and characteristics. The robot was thought to be funny when 'he' belched. He was also described to be 'outrageous' when he did not respond to an arrangement. He was seen in some instances as being shy. Although the robot has a gripping arm, this was referred to as a hand, and the red lightening bulbs in his head were designated as eyes. He was depicted as evil, as dangerous. When facing fear of the robot, insecurity was immediately triggered, causing a direct interaction with the research object. We noted sentences like: "Just go and get away from me!" One of the respondent showed fear 'for' the robot and asked: "What is hurting you now?" When the robot was in a standby mode, one declared: "Now he sleeps. We have to turn it off. He was bothered enough. He cannot afford more". Although the experimental situation did not last more than 30 min, this limited amount of time was enough to highlight how quickly quasi-interpersonal relationships between persons and robotic others can be produced. It seems that we end up using our familiar communication lexicon when dealing with robots (see also Reeves and Nass 1996).

Studies such as our own in laboratory conditions underline that the social incorporation and embedding of a robot transforms it to an 'other'—and that this robotic other or third is produced and constructed in a process of social interaction. In our case, as noted earlier, the respondents were clearly in an artificial setting and were shortly confronted with the robot. Studies performed under real life conditions and over time can provide further insights on the social incorporation and acceptance of robots and their prospects, such as experiments run in a quasi-natural setting, mainly in a reproduced home environment. However, this research and its scope remains limited and is still in its early stages (see Fernaeus et al. 2010, Forlizzi 2007, Fridin et al. 2011, Leite et al. 2013).

4.6 Concluding Remarks: Social Incorporation and Change in the Perception of Robots

The studies were conducted to examine the perception and acceptance in the context of the social representation of robots. Our research started with the idea of images of robots that give a first insight and allow one to grasp social representations. Not least, it was shown that these representations change depending on one's experience with robots, that these representations are based on communication, and finally that communication as a mutual indication is constitutive of a medium in general and of a robot more specifically. This makes a robot a meaningful object and finally a social robot.

When something becomes as a part of our everyday life and familiar to us, then the image we attribute to it changes. The mere presence of robots can lead to the establishment of such a familiarity. Whether robots are known through movies or are a major building block of one's own environment and everyday encounters makes a difference. One example is the natural experiment in Peccioli, a small city with 5,000 inhabitants in Toscana. In that context, people everyday come across robots and must react to them—even if the robots initially perform easy and limited tasks such as waste disposal. In Peccioli people's perception of and expectations from robots change as they are exposed to them in real-life conditions. Future developments will offer further opportunities for research of how robots become socially incorporated and meaningful. For instance, it could be interesting to see how people react to family robots such as Jibo, which is being developed by Cynthia Breazeal, whose name and earlier work are rather bound to the robot Kismet (2002, 2003). Jibo will be a considerable benefit to the current research to examine how a robot like this will be accepted or domesticated in a familial social environment. This will permit not only the investigation about individual acceptance of robots but also about their social acceptance. Furthermore, and finally, this will show robots are not prefabricated machines, but rather they are 'made' and 're-invented' based on what people do with them.

References

Blumer H (1986) Symbolic interactionism. Perspective and method. University of California Press, Berkeley

Breazeal C (2002) Designing sociable robots. MIT Press, Cambridge

Breazeal C (2003) Towards sociable robots. Robot Autonom Syst 42:167–175

Duffy BR (2003) Anthropomorphism and the social robot. Robot Autonom Syst 42:177–190

European Commission (2012) Special Eurobarometer/Wave EB77.1: Public attitudes towards robots. Report. http://ec.europa.eu/public_opinion/archives/ebs/ebs_382_en.pdf. Accessed 1 March 2014

Fernaeus Y et al (2010) How do you play with a robotic toy animal?: a long-term study of Pleo. In: IDC interaction design and children 2010. Barcelona, Spain, 9–12 June 2010, pp 39–48

Forlizzi J (2007) How robotic products become social products: an ethnographic study of cleaning in the house. Human-computer Interaction Institute, Carnegie Mellon University. http://

repository.cmu.edu/cgi/viewcontent.cgi?article=1037&context=hcii. Accessed 1 March 2014
Fridin M et al (2011) Acceptance, interaction, and authority of educational robots: an ethnography study of child-robot interaction. In: IEEE workshop on advanced robotics and is social impacts 2011. California, USA
Höflich JR (2013) Relationships to social robots: towards a triadic analysis of media-oriented behavior. Intervalla 1:35–48
Isbister K, Nass C (2000) Consistency of personality in interactive characters: verbal cues, non-verbal cues, and user characteristics. J Hum Comput Stud 53:251–267
Kahn PH et al (2004) Social and moral relationships with robotic others? In: Proceedings of the 2004 international workshop on robot and human interactive communication. Kurashiki, Okayama, Japan, 20–22 Sept 2004
Leite I et al (2013) Social robots for long-term interaction: a survey. Int J Soc Robot 5:291–308
Mori M (2012) The uncanny valley. IEEE Robot Autom Mag 19:98–100
Mori M (1970) The uncanny valley. Energy 7:33–35
Moscovici S (2002) Social representations. Explorations in social psychology. New York University Press, New York
Nass C, Moon Y (2000) Machines and mindlessness: social responses to computers. J Soc Issues 51:81–103
Neilsen S (2011) Robot nation. Surviving the greatest socio-economic upheaval of all time. Eridanus Press, New York
Norman D (1988) The design of everyday things. Currency Doubleday, New York
Nomura T et al (2006a) Measurement of anxiety toward robots. Interac Stud 8:347–454
Nomura T et al (2006b) Altered attitudes of people toward robots: investigation through the negative attitudes toward robot scale. American Association for Artificial Intelligence. http://www.rikou.ryukoku.ac.jp/~nomura/docs/NARS_AAAI06.pdf. Accessed 1 March 2014
Nomura T et al (2006c) Exploratory investigation into influence of negative attitudes toward robots on human-robot interaction. In: Lazinica A (ed) Mobile robots toward new applications. Pro-literatur Verlag, Augsburg, pp 215–232
Pew Research Center (2014) AI, robotics, and the future of jobs. http://www.pewinternet.org/files/2014/08/Future-of-AI-Robotics-and-Jobs.pdf
Ray C, Siegwart R (2008) What do people expect from robots? In: 2008 IEEE/RSJ international conference on intelligent robots and systems, Nice, France, 16–22 Sept 2008
Reeves B, Nass C (1996) The media equation. How people treat computers, televison and new media like real people and places. Cambridge University Press, Cambridge
Reichardt J (1978) Robots. Fact, fiction, and prediction. Penguin, New York
Simmel, Georg (1995) Soziologie. Untersuchungen über die Formen der Vergesellschaftung [Sociology. Studies on the forms of socialization]. Suhrkamp, Frankfurt
Syrdal DS et al (2009) The negative attitudes towards robot scale and reactions to robot behaviour in a live human-robot interaction study. http://uhra.herts.ac.uk/bitstream/handle/2299/9641/SyrdalDDautenhahn.pdf?sequence=1. Accessed 1 March 2014
Tsui KM et al (2011) Measuring attitudes towards telepresence robots. Int J Control Syst 16:1–11
Turkle S (2011) Alone together. Why we expect more from technology and less from each other. Basic Books, New York
Wagner W, Hayes N (eds) (2005) Everyday discourse and common sense. The theory of social representation. Palgrave, New York
Winograd T (1996) Profile: Macintosh human interface guidelines. In: Winograd T (ed) Bringing design to software. ACM Press, New York, pp 81–85
Yanco HA, Drury JL (2002) A taxonomy for human-robot interaction. In: Proceedings of the AAAI Fall Symposium on Human-Robot Interaction, pp 111–119
Zhao S (2006) Humanoid social robots as a medium of communication. New Media Soc 8:401–419

Part II
Human Interaction with Social Robots

Chapter 5
Social Robotics in Health-Care Service: The Case of Rehabilitation Programmes in Hong Kong

Pui-lam Law

This chapter deals with the use of assistive technologies in rehabilitation programmes, particularly the physically disabled in Hong Kong. The chapter argues that if the assistive technologies can only restore the biological and physical functioning of the disabled, they remain a robot only. When these technologies also manage to improve the social life of the disabled, they can turn into social robots. Following this line of thought, the chapter argues that rehabilitation programmes facilitating the use of assistive technologies in Hong Kong have difficulties in transforming assistive technologies into social robots. The chapter is concluded by further elaborating these difficulties at both the micro- and macro-level.

5.1 Introduction

This chapter critically discusses the use of assistive technologies in rehabilitation programmes in Hong Kong. The particular focus of the chapter is on physically disabled people and on the rehabilitation programmes targeted at this group. The chapter is premised on the idea that assistive technologies prescribed to a rehabilitation patient are kinds of robotic devices as they serve the purpose of restoring the physical and biological functioning of the disabled patient. To become social robots, they should also manage to serve the social functioning of the disabled patient by improving their day to day social life.

P. Law (✉)
The Hong Kong Polytechnic University, Hong Kong, China
e-mail: patrick.law@polyu.edu.hk

J. Vincent et al. (eds.), *Social Robots from a Human Perspective*,
DOI 10.1007/978-3-319-15672-9_5

Following this line of thought, the chapter analyses the rehabilitation programmes that prescribe assistive technologies for disabled people in Hong Kong. The aim is to identify some key obstacles to turning these assistive technologies into social robots. First, the chapter briefly introduces recent rehabilitation programme development in Hong Kong. Second, using wheelchairs as an example, the chapter delineates the problem of turning the wheelchair into a social robot at both the micro- and macro-level. The chapter concludes by arguing that turning an assistive technology into a social robot requires communication between the social and the technological.

The chapter data are based primarily on participant observation of wheelchair users and the interviews of professors of biomedical engineering and a frontline worker working with the disabled.[1] The professors have been teaching in rehabilitation programmes, both at the undergraduate and postgraduate levels, for years. In addition, they are also well-experienced engineers in designing and inventing assistive technologies for disabled patients. The frontline worker has been working with rehabilitative patients for many years. The framework of the interviews focuses on the assimilation of the technological into the social aspect. Detailed questions are, for instance, on evaluating the needs of patients, the prescription and design of the technologies, and structure of rehabilitation in integrating the technological with the biological and the social. While the interviews reflect the professional perspective, participant observation supplements the user perspective.

5.2 A Brief Introduction to Rehabilitation Programmes in Hong Kong

Hong Kong is a very advanced modern city in terms of its well-developed medical service and well-rounded rehabilitation programmes. In the early 1970s, Hong Kong had already started developing the notion of integrating patients with various kinds of disabilities into the community (HKSAR 1977). Since then, rehabilitation services have been growing rapidly. For instance, rehabs, day centres, halfway houses and vocational rehabilitation services were all in place to support rehab patients returning to full social functioning.

Moving on to the 1990s, the Hong Kong government issued another white paper on rehabilitation that has emphasised both the equal opportunities for the disabled and a full integration of rehab patients into the community (HKSAR 1995). Thus, new policy on barrier-free access was introduced. For instance, all public transportation should have barrier-free access facilities for the disabled with wheelchairs.

[1]Thank you to Prof. Arthur FRT. Mac, Prof. Daniel O.K. Chow, Ir Dr. Eric O.K. Tam and Mr. Ivan Su.

In the 2000s, the government introduced social enterprise. The rehabilitation sector can apply for funding from the government to set up social enterprise for rehab patients. This scheme can therefore provide more employment or training opportunities for rehab patients; this is central for development care, and self-support for enabling these patients to return to the community.

In 2005, the Rehabilitation Advisory Committee conducted a major review on the Hong Kong rehabilitation programme plan to make recommendations on rehabilitation services for future development in the following two strategic directions:

> to promote cross-sectoral collaboration in providing barrier-free environment and diversified services for persons with disabilities so as to facilitate their integration into the community; and to empower persons with disabilities and their carers, so as to help them become valuable social capital. (RAC 2007)

The review report provides a more comprehensive recommendation on future rehabilitation services in namely: pre-school training, education, employment and vocational rehabilitation, residential care, day care and community support, access and transport, development of self-help organisations, application of information and communications technologies and recreational and cultural activities. Undoubtedly, according to this report (Rehabilitation Advisory Committee (RAC) 2007), rehabilitation services in general in Hong Kong have been improving significantly for the past four decades. However, when we look into the rehabilitation services, particularly the service using assistive devices or technologies for persons with physical disabilities, we find that there is a very serious communication problem between the social and the technological aspects. This communication problem can be attributed to several factors, namely the indecisive meaning of social needs, the market supply issue, and the fragmentation of healthcare services. All these problems have contributed to obstructing the process of turning assistive technologies into social robots.

In what follows, I shall use the wheelchair as an example to elaborate the communication problem in greater detail.

5.3 Social Robotics and Assistive Technologies in Rehabilitation Programmes

Those who need to have a prescription for a wheelchair have physical disabilities concerning mobility such as a musculoskeletal or paraplegic problem, a neurological issue that mainly affects the leg or the lower part of the body, or have injured legs or broken bones in the lower part of the body. For this group of rehabilitation patients, the wheelchair is designed to solve the human mobility problem—that is moving from one place to another or obtaining something that is out of reach.

A wheelchair is indeed among the most basic assistive technologies in a rehabilitation programme. Most of the wheelchairs we see on the street are manually propelled, that is either propelled by the occupant or pushed by someone. There are also powered wheelchairs that consist of an electric motor and a simple navigation control system such as joysticks. Some are more advanced, which are controlled by voice or thought (Simpson and Levine 2002; Rehsamen et al. 2007). Although a wheelchair does not have any sensor or programmable interface directly linking the cognitive function of the person with the mechanic function of the chair itself, it can still be considered as a kind of simple and basic robot (Miller 1998).

But can a wheelchair function as a social robot? In designing assistive technology for a physically disabled patient, the engineer is simply concerned with whether the wheelchair can restore the physical or biological functioning that the patient has lost. Thus, for example, a prosthetic lower limb is designed by the biomedical engineer for an amputee to restore the mobility function. But mobility is not only limited to physical functioning; it also has its social aspects. For instance, if the amputee is Japanese, his way of sitting in a traditional Japanese restaurant would be completely different from sitting in a non-Japanese restaurant. The Japanese person has to sit in a kneeling form. The prosthetic lower limb designed for the Japanese person also has to take care of the daily social functioning of the patient. Thus, if a wheelchair could only solve physical or biological functions, then it remains a robot. But if it can perform social functions as the social robot targets in everyday life, then it is a social robot (Fortunati 2013). In a rehabilitation programme, if the patient is prescribed a wheelchair as an assistive technology to restore the social functioning of a disabled patient that has lost his or her mobility function, then the wheelchair should also serve the social function rather than merely the robotic function.

Cost-Benefit Model of Funding In a rehabilitation programme for a patient with a mobility problem, understanding the social needs of the patient is essential for the biomedical engineer in designing the wheelchair for the patient. In Hong Kong, however, the biomedical engineer always encounters frustration when dealing with the social aspect of designing the wheelchair for the rehabilitation problem. Communication is always the first problem they face.

In Hong Kong, a wheelchair can easily be bought in the market. But without doubt, they are usually very simple in terms of design, and it may not fit the needs of a particular patient. For instance, if the patient is socially active and likes to go out frequently, then a lighter wheelchair would be more convenient for travelling around. In addition, a lighter wheelchair can also minimise shoulder pain for the patient if it is manually propelled. Inevitably, this type of wheelchair would be more expensive than an ordinary one.

In Hong Kong, if a rehabilitation patient is not wealthy enough to get a lighter wheelchair, they have to go to the medical social worker to apply for financial support. Interestingly, the social worker always questions why the rehabilitation patient needs to have an expensive lighter wheelchair even if the biomedical engineer provides an explanation. Tied up by red tape, the amount of money that the patient can get usually only covers half or less than half of the price of getting a lighter wheelchair. If the socially active patient cannot obtain a wheelchair most suited to them it will perhaps hinder his or her social needs. Most likely the heavy wheelchair they have to buy will eventually be abandoned at home because it does not meet the needs of the user.

The bureaucracy involved in obtaining a wheelchair suitable for social needs reflects two issues here. The first is a money issue. In Hong Kong, the cost-benefit model has influenced social service funding policies. The principle of measuring whether a service is successful depends on the calculation of the balance between the quantity of resource input and service output. According to this principle, giving full funding to a lighter wheelchair would of course violate the cost-benefit model. The prescription of a powered wheelchair is an example of the problem of a cost-benefit model. A well-experienced therapist can easily tell that a rehabilitation patient would definitely have shoulder problems after using a manual wheelchair for years. Thus, for the long-term well-being of the patient, both the biomedical engineer and the therapist would always believe that prescribing the patient a powered wheelchair is much better than a manual one. The funding application would of course be flatly rejected, as this is contrary to the cost-benefit model. Obviously, the patient will then develop shoulder joint illness after using the manual wheelchair for years.

The Funding Model Problem The second issue is the problem of communication between the social and the biomedical. When prescribing a wheelchair to a patient, different patients of course have different social needs; therefore, the types of wheelchairs prescribed would vary according to different patients. But the biomedical engineer finds that the medical social worker has never listened to their description of the various biological and physical needs of the patient. Whatever type of prescription, the medical social service would provide funding that can only get the simplest and the most basic wheelchair model. This wheelchair example reveals in general terms that the problem of communication between the social and the biomedical has been conducive to the lack of a funding model for the prescription of suitable assistive technologies for patients.

In view of the above issues, if the wheelchair is considered as a robotic extension of the patient, it is only a poorly performed robot, and not a social robot.

Social Needs Whether a wheelchair can work as a social robot to restore social functioning of a rehabilitation patient always worries a biomedical engineer. What a biomedical engineer can do is to design a wheelchair that meets the biological, physical, and, to a very small degree, the social needs of a patient.

For instance, a wheelchair-bound patient having a sitting problem would most likely subsequently have back pain illness. A biomedical engineer can design a wheelchair that helps the patient to sit in an upright mode; but in order to solve the sitting problem, it is usually the height of the seat that has to be increased. Consequently, this would create an interaction problem of the patient with his or her living environment: The height of the washbasin or the dining table may not fit the height of the wheelchair; and furthermore, the change in the height of the wheelchair will lead to a change in the way the health-care worker moves the patient. Thus, if a wheelchair can become a social robot it should at least help the patient to interact with the living environment efficiently, otherwise, it would likely be abandoned.

In addition, if a wheelchair is to be a social robot, it should also be designed to meet the patient's social needs. Understanding the social needs of a patient frequently presents difficulties for the biomedical engineer in Hong Kong. The biomedical engineer could visit the patient's living environment but clearly a social worker's involvement is also necessary assist in linking needs with the design. However, as aforementioned, the medical social work in Hong Kong does not communicate with the biomedical engineer in developing a suitable funding scheme for a wheelchair prescription, let alone a deeper understanding of how the wheelchair can fit the social needs of the patient. This is the fragmentation of the biomedical and the social.

Adjustment Issue A simple understanding of the rehabilitation process using assistive technologies would integrate the technologies with the human body in order to restore social functioning and cope with daily social needs. It is easier for a child than an adult patient to integrate assistive technologies, and a wheelchair in particular, simply because he or she is still developing social needs. A wheelchair-bound child can play basketball instead of soccer whereas an adult patient who has already developed a way of life may find it a difficult task to adjust their established social needs and preferences with the prescribed technologies.

In addition, the rehabilitation process is much more difficult at the convalescence stage than at the acute stage. At the acute stage, the patient learns how to use the technologies at the rehabilitation centre with the help of the physiotherapist or the occupational therapist. Thus in the rehabilitation centre, the patient experiences the technology acquiring stage, but how to transfer the skills they have acquired to their daily living environment is another question. The real test of whether integration of the technologies to the rehabilitation patient's body is successful is whether the patient really knows how to use them in their real life.

During the convalescence stage, when the patient returns home, he or she has to use the technologies to interact with the living environment, both physical and social. At this stage, help from family members is very important. First of all, family members also have to learn how to use the technologies. Even with a simple wheelchair, family members have to understand the problems and limitations of the wheelchair when the patient is using it at home. When the wheelchair cannot fully fit the home environment, then home modification is necessary. In addition, either the patient takes the initiative to advocate the problem of using the wheelchair at home or family members have to be sensitive to the patient's problem. Thus, the rehabilitation process is not just how the patient learns to integrate the wheelchair with his or her body, but rather it is a holistic process involving family members who must also participate in the process of integration. In short, if a wheelchair can become a social robot is not merely an integration of the body and the machine. It is a process involving the patient and the assistance of the patient's family members working together to help the patient participate in his or her social environment.

In Hong Kong, however, the concept of community rehabilitation provides only a supply of food, home cleaning or home visit therapy. Conducting home visit therapy is the same as using the patient's home as a rehabilitation centre as the therapy provided is similar to what the patient received in the acute stage. Most rehabilitation patients cannot find success in using their technologies: Some even have two wheelchairs, one for outside and the other for home use. Without the help of family members, they would most likely retreat to their beds and their wheelchairs would be abandoned and become rusty.

Hence, integrating the wheelchair so that it becomes a social robot is also a social process. The social process can be a process provided by the community rehabilitation programme; it can also be a process, a macro social process, which a rehabilitation patient experiences when he or she returns to society.

Market Issue Sometimes a patient may not easily find a suitable wheelchair for him or herself even if he or she is wealthy enough to obtain the best in the market. The following examples illustrate that the design of a wheelchair can fit some basic social needs of a rehabilitation patient but this design cannot be easily found in the market.

The first example is about the wheelchair anti-tip device. In Hong Kong, when a patient goes to the market to buy food or other daily necessities with their domestic helper, the helper or the patient often hangs bags of stuff on the push handles on the back of the wheelchair. This phenomenon is indeed common in Hong Kong. If the bag is too heavy, it would easily cause the wheelchair to topple and the patient then gets hurt. In order to be safe, the biomedical engineer would always advise the patient acquires a wheelchair with an anti-tip device. Interestingly, it is hard to obtain a wheelchair with an anti-tip device in the Hong Kong market, the reason being simply because it is inconvenient for the helper or the one pushing the wheelchair.

Some assistive technologies are also considered orphan technologies. In Hong Kong, prescribing a wheelchair for a rehabilitation patient always encounters the problem of a high abandonment rate. One of the many reasons is that the living environment is too small for using a wheelchair. But even if a wheelchair can be designed with movement in 360° directions, apparently, as demand is limited, it is difficult to attract large-scale production. This issue of lack of market scale demand means it is generally difficult to find a suitable wheelchair that fits the living environment, in addition to assistive technologies in general.

Community Problem As early as 1985, Hong Kong already had a Building Ordinance (The Hong Kong Special Administrative Region (HKSAR) 2012) requiring various kinds of buildings to have barrier-free access for disabled persons. In 1995, the Disability Discrimination Ordinance (HKSAR 2013) was enacted to protect disabled persons from unfair treatment. Moving on to the 2000s, rehabilitation programmes have emphasized helping the patient return to the community. Despite the fact that the government strongly supports community-based rehabilitation programmes, it seems that people's understanding of the notion of rehabilitation in Hong Kong in general is still very weak. The production of assistive technologies is a good example.

As already noted, most assistive technologies are considered as orphan technologies because demand is limited; for example, the wheelchair with an anti-dip. But these technologies should not be purchased individually in the general market place if they serve the basic needs of a rehabilitation patient, rather it should be supported by the community (Walzer 1983). However, in Hong Kong, even the funding model of a wheelchair is not primarily developed for the well-being of a rehabilitation patient, let alone our community's understanding of the social needs of the patient.

First, Hong Kong is a modern and vibrant city; the pace of the city is fast. We are seldom aware of what is happening around us. Although we have the hardware such as a rehab bus or barrier-free environment for a disabled person, we may not notice that a disabled person is walking next to us. Our fast pace may blind our sensitivity to barrier-free access facilities and the wheelchair patient who may need space to move on the road: some areas are too congested because pedestrians use the wheelchair pathway, resulting in blocking the road for the wheelchair patient. Thus, the atmosphere of the city itself is detrimental to the idea of bringing the disabled back to the community.

Second, as we have learned thus far, some of the hardware provided for the disabled may not really meet their social needs. Many Hong Kong people, particularly the elder ones have the habit of going to a Chinese restaurant every morning to enjoy tea and *dim sum* (a type of Cantonese food for breakfast). However, there is always a problem for the wheelchair rehabilitation patient to go from their home to their nearby tea restaurant, simply because the restaurant may not have any barrier-free access or the route is not suitable for wheelchair transit. This would of course limit their social recreational activities, which the community rehabilitation programme has strongly promoted (Cheung et al. 2009; Welage and Liu 2011).

Another example is the organised wheelchair marathon competition by the Standard Chartered Hong Kong Marathon (the Marathon). The Marathon introduced a full and a 3-km marathon wheelchair competition in 2012. But the full marathon was indeed too long a distance for the wheelchair participants and was cancelled in the following years. In 2014, in addition to the 3-km marathon, the Marathon piloted a 10-km wheelchair marathon in order to promote recreational activities for the disabled. Because the wheelchair competitors and the runners cover the same route, which includes very steep inclines, it was too difficult for the wheelchair competitors to finish the 10-km race. Subsequently, the Marathon re-routed the 10-km wheelchair race for the upcoming 2015 marathon. Despite the fact that the organising party really wants to promote the well-being of the disabled, this example, however, illustrates that they are not sufficiently aware of the problems of wheelchair competitors.

It has become clear that Hong Kong is not only weak in the community rehabilitation programmes it offers, but that it also lacks a macro framework of community rehabilitation. The purpose of the community rehabilitation programme is to help patients return, as much as possible, to their social life. These programmes focus on bringing the patient back to their family, or providing service in day centres, halfway houses and vocational training, or recently the establishment of social enterprise. All these programmes are, however, patient centred, that means focusing on how to empower patients to adapt themselves to the community. As mentioned in the previous section, when a patient is discharged and returns home, the rehabilitation programme would not be effective if family members do not or are unable to participate together in the programme at home. And by the same token, when the patient returns to the community at large, the community should also participate together to work with the patient, helping them to return to their normal social functioning.

Considering the example of going to Chinese tea restaurants; some wheelchair patients, by means of using social media, have organised a self-help group to identify and then inform those who are fond of going to tea restaurants the best possible wheelchair accessible route from their home to the most nearby restaurant. But these groups are organised by patients themselves and hence may not be able to cover more information or provide support for the social needs of each of the patients. If the government could take the initiative to organise this kind of group, it would help patients to adapt back into the community more easily on the one hand, and to arouse the community's awareness in helping patients using assistive technologies to participate in the social environment at large, on the other. A wheelchair patient may have already gained the capacity to deal with the physical and social environment at large; however, support from the community is also central for their participation (Chan and Chan 2007).

5.4 Wheelchairs and Social Robotics—Concluding Remarks

Whether a wheelchair is a social robot or not is indeed inherently related to the successfulness of the rehabilitation programme. In view of the Hong Kong rehabilitation programmes developed for wheelchair patients at the present stage, it seems that wheelchairs can only function as robots rather than social robots. The issue has been analysed at two levels, namely the micro- and the macro-levels.

At the micro-level, it is all about how the patient integrates him or herself with the wheelchair. The integration is of course not merely fitting the wheelchair to the patient from a physical or biological aspect. The purpose of the integration is to restore to the greatest extent the social functioning of the disabled patient. Integrating the biological and physical aspects are only the necessary conditions for the restoration of the social functioning. This usually takes place at rehabilitation centres or hospitals in which the patient learns how to integrate the assistive technologies with their body. This is the stage of capacity training.

In the second stage when the wheelchair patient returns home, the patient's capacity for using assistive technologies in participating in the environment is tested, both physical and social. Thus this stage is the negotiation between the assistive technologies and the environment, particularly the social environment. In Hong Kong, it is often at this stage that the patient faces frustration, as the wheelchair cannot help them participate in their routine social life. The Hong Kong rehabilitation programme plan suggests a cross-sectional collaboration, while in the actual situation there is always a problem of communication between the social and the technological. Lacking communication between the social and technological, the patient and the patient's family members can barely find the meaning of the wheelchair in their social life. The wheelchair remains a technology or a robot; it can hardly become a social robot.

Despite the fact that some wheelchair patients can participate in the social environment at home, they may encounter frustration when they return to the community at large. In Hong Kong, the government has been enhancing barrier-free access and transport for the disabled through facilities that provide equal opportunities for integrating into the community. Undoubtedly, this hardware is built to fit the mobility of the wheelchair, but whether it is fit for the patient is another question. At very least we can see the disadvantage at the recreation or sports levels where the government has deliberately developed their programme plan. The problem may be attributed to social issues. Our community has been developing very good hardware for disabled persons such as the wheelchair patient. Our community at large, however, lacks a good understanding, sensitivity or patience towards the patients. The steep inclines of the 10-km wheelchair competition is an obvious example. Similar to the micro-level, having the hardware for the wheelchair is only a necessary condition for the patient returning to the community at large at the macro-level. But how to implement and integrate the hardware so that the wheelchair can be part of the community at large requires equally a dialogue between the social and the technical—the bio medical engineers, the social workers, the Hong Kong authorities, the patients and their families. Indeed all those involved in the everyday life of the rehabilitation of disabled people. At the macro-level, a dialogue between the social and technical is the sufficient condition for turning both the wheelchair and the barrier-free hardware into a social robot; otherwise, it remains a robot if it only serves the function of mobility.

References

Chan SC, Chan AP (2007) User satisfaction, community participation and quality of life among Chinese wheelchair users with spinal cord injury: a preliminary study. Occup Ther Int 14:123–143

Cheung MC, Ting W, Chan LY, Ho KS, Chan WM (2009) Leisure participation and health-related quality of life of community-dwelling elders in Hong Kong. Asian J Gerontol Geriatr 4(1):15–23

Fortunati L (2013) Afterword: robot conceptualizations between continuity and innovation. Intervalla Platform Intellect Exch 1:116–129

Miller DP (1998) Assistive robotics: an overview. In: Mittal VO, Yanco HA, Aronis J, Simpson R (eds) Assistive technology and artificial intelligence: applications in robotics, user interface and natural language processing. Springer, Berlin, pp 126–136

Rehabilitation Advisory Committee (RAC) (2007) Hong Kong rehabilitation programme plan report. http://www.lwb.gov.hk/eng/advisory/rac/rpp_report.htm

Rehsamen B, Teo CL, Zheng Q, Ang MH Jr (2007) Controlling a wheelchair indoors using thought. IEEE Intell Syst 22(2):18–24

Simpson RC, Levine SP (2002) White paper on integrating the disabled into the community: a united effort. Hong Kong Government Printer, Hong Kong

The Hong Kong Special Administrative Region (HKSAR) (1995) White paper on equal opportunities and full participation: a better tomorrow for all. Hong Kong Government Printer, Hong Kong

The Hong Kong Special Administrative Region (HKSAR) (2012 [1956]) Building ordinance, Gazette Cap. 123. Hong Kong Government Printer, Hong Kong

The Hong Kong Special Administrative Region (HKSAR) (2013 [1995]) Disability discrimination ordinance, Gazette Cap. 487. Hong Kong Government Printer, Hong Kong

Walzer M (1983) Spheres of justice: a defense of pluralism and equality. Basic Books, New York

Welage N, Liu K (2011) Wheelchair accessibility of public buildings: a review of the literature. Disabil Rehabil Assistive Technol 6(1):1–9

Chapter 6
Intuitive Interaction Between Humans and Robots in Work Functions at Industrial Environments: The Role of Social Robotics

Antonio B. Moniz

The social dimension of worker–robot interaction in industry is becoming a decisive aspect of robotics development. Many problems and difficulties of robotics research are not only related to technical issues but are framed by social aspects. Human–robot interaction (HRI) as a specific research field of robotics tackles this issue of intuition. One of the aims is to identify relevant research questions about the possibility of the development of safer robot systems in closer human–machine intuitive interaction systems at the manufacturing shop floor level. This chapter will contribute to understanding the cognitive and perceptual workload for robot operators in complex working systems. The importance of robotics in work life is not only to decrease the physical strains in manufacturing, but also it can increase the need for situation awareness and risk assessment which implies higher perceptual workload and psychological strains. The social sciences approach to such technology assessment is of high relevance in order to acknowledge the dimension of the intuitive interaction concept within social robotics.

6.1 Introduction

In recent debates it has become important to understand the definitions of social robots' abilities when they can (or not) be applied to "companion" robots. But the discussion on robots with interaction capabilities in work environments has not been included under the topic of "social robotics", rather, this definition has been

A.B. Moniz (✉)
CICS.NOVA, FCT-UNL, Universidade Nova Lisboa, 2829-516 Caparica, Portugal
e-mail: antonio.moniz@kit.edu

A.B. Moniz
ITAS, Karlsruhe Institute of Technology, Karlsruhe, Germany

J. Vincent et al. (eds.), *Social Robots from a Human Perspective*,
DOI 10.1007/978-3-319-15672-9_6

applied to some type of so-called "cobots" (Colgate et al. 1996). These include 'social interaction with robots' features (perception, sensing, haptic interaction, communication) where people interact with robots that have some degree of awareness of the human in terms of sensing abilities and/or interfaces and abilities to interact and communicate with people. The fact that robots with such abilities are introduced in a working environment means the relation between humans and these machines also evokes the relation between co-workers and the human resource management strategies in a company (Moniz 2014). It relates to a problem of human–machine interaction, and the complexity of the work environment (Bijker et al. 1987). It also becomes a dimension of job design where technical and social criteria must be taken into account for the design of tasks and for the communication process (Huws 2006; Weiss et al. 2009; Bernstein et al. 2007). Meanwhile, in industry there are a lot of examples of machine operators on the shop floor that interact with automated systems using sophisticated communication features. These can include not only data management (robot programme changes, data input and retrieval) but also oral communication features. A robot which is assumed to have 'social' ability will require the ability to perceive its environment (*perception*) and to reason about it (*cognition*), likely including the ability to detect social cues and to reason about the world from the perspective of others. With this in mind, we can also say that the new generation of industrial robots can also be recognised as "robots capable of social ability". But in these cases, the sound communication is not as relevant as is the visual one. This implies that the definition of "language" should be considered as well as gestural communication. In any case, such applications can use "natural interaction" as their most important elements. Nevertheless, the communication capacities are not sufficient to classify these abilities as "social".

Perception has become possible through the use of advanced sensor integration, which can be useful for the human operator by providing information where humans have difficulty such as in collecting data. The cognitive feature is the most difficult one as it could be useful for the human operator when presenting different alternatives during problem-solving. Solutions such as this can be built upon the operator inputs and from knowledge databases that such robots can use. As it is usual for robot cells to be operated by different humans (in the same working group or in different shifts), the knowledge management or reasoning can become useful for task performance and for problem-solving.

The minimization of the cognitive and perceptual workload for robot operators in complex working systems is very important, because it interferes with the task performance and with operational safety. That can be highly relevant when different robots with different roles and different designs are to be used in the manufacturing industry to a larger extent. It is also necessary to investigate the transferability of results from industrial environments to other fields where the introduction of robotics is planned such as in health care, agriculture, mining, underwater, logistics, space operations, inspection, disaster management, medicine and so forth. This chapter examines in the following four sections the role of social robots from the perspective of complex environments, intuitive interaction, cobots, sharing workspace and concluding with some final remarks including consideration of safety issues.

6.2 Working with Robots in Complex Environments

Starting from the conceptualisation of 'intuitive interaction with technology' of robotic systems I aim to discuss applications within industrial environments using the 'social' robotics approach. Such a concept should not be only applied to humanoid systems. Complex working environments (CWE) implies the interaction of humans with automated systems, and more often they include robots. This has increased the possibility for eventual malfunctions or even dysfunctions; when they occur the impacts can be severe.

During the 1980s, it was said, for example by J.F. Bard that "in general, robots should be capable of outperforming a person in hostile working environments where noise, vibrations, toxic fumes and other insults are present. Nevertheless, they cannot operate in a disorderly setting. Parts to be handled or formed must be in a known place and have a known orientation" (Bard 1986, p. 102). This means that even in an unstructured work environment which has been called by Bard a 'disorderly setting' a robot cannot operate. There have been important technical and conceptional developments since then, but this assumption continues to be true. A balance should be found between the need to use a robot in aggressive tasks of repetitive action or in hostile environments, and the lower capacity of a robot when compared with the human performance.

In many sectors, such as automobile, electronics and metal engineering, robots have been comprehensively introduced in this way. This means that, according to the International Federation of Robotics (IFR), around 4 million workers around the world have a close connection to robot operation in their work environment (IFR 2013). This has increased the need to consider the social dimension of interaction with technology in these environments. Ergonomics studies became fundamental in all sites where robotic systems had been introduced, but psycho- and sociological inputs are only just beginning in these areas. Indeed, most companies do not have such social scientists on their staff. The solution to close working with robots found by companies is to physically separate the robot cells from the human presence through fences or guards. This can be done without problem in larger companies but in small- and medium-sized enterprises (SME) it becomes a problem due to the lack of available space on-site. Robot manufacturers have started to develop new sensing systems and mechanical and material features that can allow the closer interaction without barriers. Solutions are still under research to develop more "intelligent" systems that integrate such sensor components and allow a more intuitive communication and interaction with humans. New robots with "social" abilities including more complex communication and reasoning capabilities will become more common in manufacturing environments.

Although there is an increased use of social robots for industrial sites, there are still important features that can only be done by human operators. In answer to the question what are the main task roles in a complex working environment one can find the following:

- Operation control
- Maintenance

- Operation monitoring
- Quality control

The performance of tasks must be understood within context of the compliance of the aims of the task in a chain or system of tasks, and the features of its performer (a human or a machine). The question being asked here is: will a human be replaced by a robot? Perhaps even by a robot with increased intelligence and social abilities? My answer, based on my research experience and by literature interpretation, would be no. That answer is also based on the fact that, whenever the more "intelligent" the automatic system of machines becomes, the more complex the problems that will occur. Anyway, the task roles attributed to humans and machines in work environments must be analysed according to all sets of conditions. When tasks are not designed according to the attributes of the performer, the outcomes will not be those that are usually expected. This can happen in automated or in conventional operating systems. Some malfunctions can occur, or even accidents. In other words, "unexpected events" may occur.

To run a batch manufacturing shop on an around-the-clock basis, systems have to be able to respond to unexpected events, such as extra stock, defective material, and premature tool wear out. But Bard added also a curious statement: "Adaptive control, coupled with robots, makes this possible by largely eliminating the need for a skilled operator to be present" (Bard 1986, p. 103). This is one of our key issues. To be precise, it seems that whenever the working environments are more complex or dense, the less it will be possible to "eliminate the need for a skilled operator".

Some authors also point to the "system responses" which lead to a specific behaviour. But how do systems respond? Do they react, or are they providing information?

In the white paper from the EURON Special Interest Group on Cooperative Robotics published in 2008 it was anticipated that for the next 10 years there would be an advance of "high-level cooperative cognitive skills, while there is a substantial need for improvement of individual cognitive skills, the ability to achieve cooperation in planning, decision making and environment modeling is the key to the development of network robot systems (NRS)" (Saffiotti and Lima 2008, p. 8). To understand this statement one should not translate those "cognitive skills" as being applied just to machines. It would be too naïve to expect such autonomous capacity. It makes sense now when we understand it applied to the interaction with humans. In the same document and about the same expectations, the group also discusses the HRI, stating that:

> Better interfaces to control and interact with NRS will improve usability and make new, broader applications possible. On the one hand, improved distributed cooperative perception capabilities of NRS will make it possible to have effective interaction with people, by understanding different kinds of signals coming from single and multiple persons sharing the NRS space; on the other hand, a scenario with multiple users interacting with multiple robots brings about new challenges that will significantly impact on HRI (Saffiotti and Lima 2008, p. 8).

Probably, the most important question now is to ask "who makes the final decision?". An answer to this question will enable us to understand how those "systems" are organised.

In general, in the manufacturing production based on automated equipment, the fact that "unexpected events" can occur gains especial importance because accidents, malfunctions or disturbances would impact the working conditions, the expected productivity and all the outcomes. Disturbances with conventional systems are usual and are considered as a cost controllable element, but with automated systems, each time unit without production represents a much higher cost. The production volume per time unit with automation is much higher than with conventional equipment and when those "unexpected events" occur in an automated environment the implications for the economic efficiency (costs, delivery times, quality) are not negligible.

To understand this dimension consider the fact that when skilled operators are taken out of complex production systems it can lead to increased failures and accidents. The implications of those decisions on economic efficiency are at stake; decreased labour costs in an organisation can mean an increased probability of disturbances or "unexpected events", which can become a risk factor. Thus, this can be one of our key issues to be discussed: job displacement and knowledge use.

Another way of exploring these problems further is to answer the following question: are the CWE trustworthy without skilled and responsible workers involved directly? If the answer is positive, that would mean intelligent non-human agents are enough to govern those environments. It would also mean humans should rely on autonomous technology in important decision processes. However, if the answer is negative, a responsible and precautious principle would be to advise humans are always included in the loop. The more complex the working environment would be, the more important it is to involve humans. This assumption brings again the qualification, training and education elements; they become crucial to understanding the problem.

We can also ask if it is possible to develop CWE with unskilled labour? This would mean that in spite of the complexity of the working environment, the qualification is not meaningful. The problem arises when one characterises "complexity". If by complexity we mean just the interconnection of several sets of equipment with some degrees of complexity, but with a high degree of automation, one can conclude there is the possibility to integrate less skilled labour in such environments. These workers could have only minor controlling or monitoring functions and this occurs in several cases, in particular, in larger companies. The problem is that for some "unexpected events" there is no capacity to solve the incident in the minimum possible time. Usually, such occurrences start a complex and large process of decision-making and demands for external experts (technicians, engineers, etc.).

We can conclude that unskilled jobs are better applicable in simpler working environments. Those that require more complex task content also need higher levels of labour qualification. It is easily observable that more complex technologies require tasks with complex contents, and this in turn always demands higher levels of skills and qualifications. Those tasks are usually related to monitoring, controlling, but also require capabilities of fine-tuning programming and maintenance. Operators with those capacities are also able to get more involved in the decision process and in the governance processes of such technology systems.

When we ask if automated systems are "unmanned" systems, what then could be the answer? The correlation would appear to be obvious but it is not supportable. An automated task does not mean that a human should not be present to assist or to be assisted. The cases where fewer humans are present in automated production systems are those in the process industry, but very few can be found in the discrete products manufacturing industry. Thus, the type of production can be a factor that influences the possibility of human involvement in the transformation process.

Finally, which implications for "unexpected events" can reveal the work function in manufacturing? Answering this question means that with the development of more complex production systems, the probability of "unexpected events" occurring is higher. They are especially higher when the systems become unmanned, that is without human control, and is why the prevention of "unexpected events" needs the inclusion of humans in the production process; there becomes a clear "work function". Once there is no work without humans the need to include humans in the automation loop implies the existence of a work function which can be for operation, for monitoring, for control, for maintenance, for programming, for tooling, or for other types of tasks that cannot become fully automated. This means that such working tasks performed by humans must include the capacity of preventing "unexpected events", or in other terms, malfunctions, or even accidents. For these reasons, it becomes so important to think about and design automated systems that necessarily include humans in the loop. Their exclusion can be understandable by a nonconformity with basic management principles. Usually, these type of organisational dysfunctions imply continuous problems in the task performance and in the productivity outputs. They imply also social distrust towards technological developments or even towards innovation policies.

6.3 Intuitive Interaction with Robots

The problems mentioned above, like dysfunctions, accidents, and other unexpected events, can be more relevant in the case of robots used in manufacturing environments. As such technology tends to become more sophisticated; even in manufacturing industry the implications for their use are becoming more important while a high volume of automated systems are in operation worldwide. That means the task roles become critical: the qualification for the job must be a factor of system performance, the capacity for programming, controlling and operation becomes even more precise, and overall the intensity of the task increases with the complexity. Great efforts have been made in order to ensure the capacity can deal with such demands. Furthermore, all the operations with most industrial robots became simpler and the interfaces became lighter and easier to use.

The study of applications on industrial environments using robots includes the arguments of intuitive interaction with technology. In a similar direction, the social dimension of worker–robot interaction is becoming a decisive aspect of robotics development. This dimension includes the knowledge necessary to operate machines and systems of machines. It is no longer just a technology problem or a technical challenge but one that is now highly relevant in CWE (robots, autonomous systems, etc.) in the manufacturing industry.

It is also necessary to investigate the transferability of results from industrial environments to other fields where the introduction of robotics is planned (health care, agriculture, mining, underwater, logistics, space operations, inspection, disaster management, medicine, etc.). Such types of new application are not only developed to increase the performance of industrial robots when those developments can also be reapplied with innovations to the traditional robotic systems, but they also became a general issue for all type of robots, including also the professional service robotics. Functions like manipulation, monitoring sensing or vision have been developed by industrial robots and now they are applied in advanced professional service robots.

However, our focus is the type of robots that have been used in work environments which until now have demanded a more or less intensive interaction with human operators. Some technological innovations have been tried in robots that act with a high grade of autonomy or without direct human interference. However, those robots that imply a common workspace with humans can present further technical challenges. The communication features have to be improved, but also all the robotic movement possibilities may interfere with the space where humans have to stand for their work environment. Such interference may cause safety problems and have to be cautiously considered in the programming phase.

The study of robotic applications and their social implications provided clear evidence of this transferability. The main research questions are usually related to industrial applications; now they can also be applied to new types of applications.

Equipped with general information about social behaviour, a robot should be able to detect situations in which certain classes of social behaviours are appropriate and to apply them. Such capacity implies also the feature of intuition in the interaction with humans. In this case, a robot can have an autonomous "reasoning" about how best to achieve its goals in a given social context, and should have the ability to express itself in ways that will help it complete tasks in a wide range of social situations. The frames of goal achievement must be settled in work environments. The higher the capacity is for "autonomous reasoning", the higher must be the intuition for humans to interact with robots. In this situation, a robot can contextualise its messages about its internal representations at this level, and "injects" these communications into the interaction in a "socially acceptable way", according to the MAR definition (MAR 2014). From our point of view, this "socially acceptable way" must be defined in a negotiated way with the working social partners, or at least with the human operators that are working with this type of robots.

6.4 Social Robots and Cobots

Recent development of robotics has enabled the emergence of the new concept of social robots as cobots. Although they do not have the same meaning they can be used in similar ways in manufacturing environments, furthermore, it is notable that when one mentions "robotic assist system", it is not only the case of health care examples that we refer to but the concept can also be applied to manufacturing operations.

Cobots are potentially well suited to safety-critical tasks such as surgery and micro-assembly, or those which involve large and powerful interaction force such as automobile assembly (Colgate et al. 1996, p. 433). Cobots are usually considered for a role as helping humans in their operative tasks and not to replace them. This point is important to state because the aim is not the accomplishment of a task with full autonomy, but the coordination of tasks with human operators, thus the interaction features are crucial.

This new technology also created particular approaches to the concept of interactive learning and safety systems of assistive robot. The traditional interactive learning with such system needs to be done on the job and most examples require learning-by-doing procedures, although other examples can emerge. The same applies to the safety measures. Assistive robotics and cobots in general imply that the equipment must operate very close to the human operator in order that he or she can be assisted. Safety rules and procedures can be strict, but those measures have to be included in the design process, and also they must involve the human operator to give information to obtain the best possible results.

6.5 Shared Workspace of Human and Robot

As we have explored in the discussion thus far operating a robot, or working together with a robot, means that humans have to share a common space. For safety reasons, a shared workspace between a human and a robot must be considered as a risk factor. Also, "a careful design of so-called intelligent assist systems (IAS) or intelligent automation devices (IAD) and their operating procedures is necessary when physical collaboration between machines and human workers also have to follow ergonomic targets" (Krüger et al. 2009, p. 628).

Sharing a workspace means that the work process must take into consideration the safety areas around robots. Interference between workspaces can occur but only when the robot is switched off thus to ensure safety, the workspaces of humans and robots are strictly separated in time or in space (Lenz et al. 2008). That implies an increased possibility for positioning the human operator with further monitoring tasks without direct intervention during operation. Under such conditions, it is difficult to consider usual robots as co-workers.

The new research developments try to overcome these limitations but to do so the consideration about safety conditions for operation is crucial. "The desired coexistence of robotic systems and humans in the same physical domain, by sharing the same workspace and cooperating in a physical manner, poses the very fundamental problem of ensuring safety for the user and the robot" (Krüger et al. 2009, p. 633). In such environments the control of operation can present limitations, and there is a need for sensor-based surveillance of the workspace.

6.6 Concluding Remarks: Safety Is Still a Key Issue

It is important to identify relevant research questions about the possibility of development of safer robot systems in closer human–machine intuitive interaction systems at the manufacturing shop floor level. As I have presented in this chapter, the features of industrial robots have been applied to service robotics and here the developments produced a whole set of innovations such as the increased capacity of human–machine interaction and communication. The autonomy features in professional service robots have enabled new developments on autonomous perception of environments. These developments could even provide autonomous "reasoning" about how the robot can achieve its goals in a given social context. Those new capacities are now applied to manufacturing robotics where the need to interact with humans is very important. However, that interaction implies a further need to focus on the safety issues when designing a production system with robots. As the complexity of work environments increases it can produce the emergence of "unexpected events" where the role of human control becomes more central.

Many authors agree that in the case of physically interacting robot assistants it is obvious that a proven safety standard is of paramount importance (Hägele et al. 2002). But safety is not only a technical feature. Anticipating possible problems or "unexpected events" is mostly a social capability that machines (and in this case, robots) cannot have. In fact, tacit knowledge, qualified and experienced jobs are key elements to ensure and improve safer workplaces with complex environments. Social robotics cannot replace those human workplaces. Robotic manufacturers are developing new safe robots to enable working alongside each other (Wallhoff et al. 2010) that would mean systems with intuitive interaction capacities to ease the co-working feature of those robots. Social robots with higher capacities of interaction and communication have the capacity to become the systems that can fit better into workplaces where human operators perform their tasks. The challenge would be how to also include these social robots in the manufacturing environments.

References

Bard JF (1986) An assessment of industrial robots: capabilities, economics, and impacts. J Oper Manag 2(6):99–124

Bernstein D, Crowley K, Nourbakhsh I (2007) Working with a robot: exploring relationship potential in human–robot systems. Interact Stud 8(3):465–482

Bijker WE, Hughes TP, Pinch T (1987) The social construction of technological systems. MIT Press, Cambridge

Colgate J, Wannasuphoprasit W, Peshkin M (1996) Cobots: robots for collaboration with human operators. In: Proceedings of the international mechanical engineering congress and exhibition (GA DSC), vol 58, pp 433–439

Hägele M, Schaaf W, Helms E (2002) Robot assistants at manual workplaces: effective co-operation and safety aspects. In: Proceedings of the 33rd international symposium on robotics (ISR 2002), p 6

Huws U (ed) (2006) The transformation of work in a global knowledge economy: towards a conceptual framework. WORKS, Leuven, p 249

IFR (2013) World Robotics 2013, International Federation of Robotics

Krüger J, Lien T, Verl A (2009) Cooperation of human and machines in assembly lines. CIRP Ann Manuf Technol 2(58):628–646

Lenz C et al (2008) Joint-action for humans and industrial robots for assembly tasks. In: Proceeding of the IEEE international symposium on robot and human interactive communication (RO-MAN 2008), pp 130–135

MAR (2014) Robotics 2020 multi-annual roadmap for robotics in Europe, EUrobotics, p 183

Moniz A (2014) Organisational challenges of human–robot interaction systems in industry: human resources implications. In: Machado C, Davim JP (eds) Human resource management and technological challenges: management and industrial engineering. Springer, pp 123–131

Saffiotti A, Lima P (eds) (2008) Two "hot issues" in cooperative robotics: network robot systems, and formal models and methods for cooperation. European Robotics Research Network (EURON)

Wallhoff F et al (2010) A skill-based approach towards hybrid assembly. Adv Eng Inform 24:329–339

Weiss A, Wurhofer D, Lankes M, Tscheligi M (2009) Autonomous vs. tele-operated: how people perceive human-robot collaboration with HRP-2. In: HRI'09: proceedings of the 4th ACM/IEEE international conference on human robot interaction, ACM. New York, pp 257–258

Chapter 7
Minimizing the Human? Functional Reductions of Complexity in Social Robotics and Their Cybernetic Heritage

Timo Kaerlein

Recently designers at the Hiroshi Ishiguro Laboratory (ATR, Kyoto) decided to shift their focus away from highly-lifelike androids to very minimalistic appearances like that of the Telenoid, Elfoid and Hugvie models. Instead of trying to simulate the appearance and behavior of actual human beings, the decision was to sidestep the Uncanny Valley problem by restricting the robots to a functional minimum in human likeness. This chapter investigates different approaches to complexity reduction in social robotics, among them experiments that compared human mime artists (theatrical robots) with actual robots in the interaction with autistic children and the implementation of cartoon techniques to model emotions in the generation of projected facial expressions. The notion of complexity reduction is discussed with respect to its merits (especially from a systems theoretical framework) and to its drawbacks. Beyond ethical considerations, the chapter argues that the research on social robotics is linked with the cybernetic hypothesis (as formulated by Tiqqun), a contemporary mode of governance and regulation that privileges certain kinds of subjectification via communication and discourages others.

7.1 Introduction

> He had bought a large map representing the sea,
>   Without the least vestige of land:
> And the crew were much pleased when they found it to be
>   A map they all could understand.
>
> (Lewis Carroll, *The Hunting of the Snark*, 1876)

T. Kaerlein (✉)
Department of Media Studies, University of Paderborn, Paderborn, Germany
e-mail: timo.kaerlein@uni-paderborn.de

J. Vincent et al. (eds.), *Social Robots from a Human Perspective*,
DOI 10.1007/978-3-319-15672-9_7

Fig. 7.1 Hugvie was developed by ATR Hiroshi Ishiguro Laboratory

The most recent telepresence robot model developed at the Hiroshi Ishiguro Laboratory (HIL, part of the Advanced Telecommunications Research Institute International, ATR, Kyoto), called *Hugvie*, resembles a soft cushion with a vaguely human shape (ATR 2012). It is less than a meter tall, comes in different colors, is made from a stretch fabric filled with micro foam beads and—most notably—has a pocket at the side of its head where users can insert their mobile phones. *Hugvie* contains a microcomputer controlling an internal vibration motor that is designed to simulate the interlocutor's heartbeat in synchronization with their voice transmitted via the inserted phone (Fig. 7.1).

Continuing the line of robotics research pursued with the *Telenoid* (2010) and *Elfoid* (2011) models before, *Hugvie* is the current manifestation of an attempt to enrich a tele-mediated conversation with some sort of haptic stimuli. Its drastically minimalistic design—remote humanlikeness, implied limbs, and general simplified anthropomorphic appearance—is a sharp shift away from the highly realistic *Geminoid* models developed at HIL (Fig. 7.2). To draw on an analogy from the area of acoustic reproduction: Whereas *Geminoid* could be described as a high

Fig. 7.2 Geminoid HI-2 was developed by ATR Hiroshi Ishiguro Laboratory. (Photo by Makoto Ishida)

fidelity system with minimal distortions due to noise, *Hugvie* on the other hand is a compressed standard of humanlikeness, delivering just the bare minimum of recognizably 'human' but de-individualized features to allow anyone interacting with the device to 'fill in the blanks.' It is part of a series of "[s]tudies on cellphone-type tele-operated androids transmitting human presence" (JST CREST 2010) and thus intended as a means to enhance existing mobile telecommunications with robotic technology. In the words of its inventors: "'Hugvie' is an epoch-making communication medium that can strongly transfer the presence of an interaction partner despite its simple shape" (ATR 2012).

The robots at HIL were chosen as the opening example of this chapter because they signify, in a rather drastic way, a tendency toward simpler shapes and less lifelike appearances in robotics design that can be encountered at various other sites too. This chapter investigates functional reductions of complexity in several areas of social robotics, among them experiments that compared human mime artists (theatrical robots) with actual robots in the interaction with autistic children (Robins et al. 2006) and the implementation of cartoon techniques to model emotions in the generation of projected facial expressions (Tsuruda et al. 2013). The notion of complexity reduction will then be discussed with respect to its merits (especially from a systems theoretical framework) and to its drawbacks. Beyond ethical considerations, the chapter argues that the research on social robotics is linked with the cybernetic hypothesis (Tiqqun 2001), a contemporary mode of governance and regulation that privileges certain kinds of subjectification via communication and discourages others.

7.2 Types of Complexity Reduction in Social Robotics

What we are dealing with in the case of *Hugvie* is first of all a reduction of complexity referring to visual traits, i.e., the abstraction of appearance to a general humanoid shape applicable in all kinds of scenarios. Instead of trying to simulate the appearance and behavior of actual human beings, the decision was to sidestep the Uncanny Valley problem (Mori 1970) by restricting the robots to a functional minimum in human likeness.[1] Furthermore, the robots at HIL are rather simple machines, they do not follow scripted behavioral patterns and feature no algorithmic social intelligence like other social robots, e.g., KASPAR (Dautenhahn et al. 2009), Paro (Wada and Shibata 2006) or Kismet (Breazeal 2002). Their main purpose is to function as media of telecommunication with some robotic functionality like motor control and humanoid shape. But of course, complexity reduction is an issue in social robotics in general. Two main types can be differentiated:

1. quantitative limitation of possible input parameters via available sensors and algorithmic modeling (what counts as a stimulus that the robot reacts to and what is ignored?);
2. quantitative (and often qualitative) limitation of possible internal states and outputs according to the robot's capabilities and use scenario.

The two types of complexity reduction are connected insofar social robots are often programmed to display a certain behavior or enter a certain state in reaction to environmental stimuli. The first type refers to how the robot perceives its surroundings (*inputs*), whereas the second type applies to its interactions with (mostly human) counterparts (*outputs*).

An important caveat is in order here: Reduction of complexity is not a defect of robotics exclusively. To be more precise, it is not even a defect at all. Niklas Luhmann, the founder of social systems theory, differentiated "two types of systems: those in which each element can be related to every other element and those in which this is no longer the case" (Knodt 1995, p. XVII). The latter provides his definition of complexity as the totality of all possible events where there are always more possibilities of action than can ever be actualized. Only a limited amount of all available information is ever selected for internal processing which is a necessary condition for systems of any kind (biological, psychic, social, etc.) to function. The ensuing complexity differential (*Komplexitätsgefälle*) between system and environment serves as a protective mechanism—if too much complexity is allowed into a system, it usually breaks down, as in the case of psychopathology. The same is true for social interactions—not every possible communicative link can be actualized in a given situation, not every subtlety of talk considered. Social systems, according to Luhmann, interface between the abundant complexity of the world and the limited resources available to face it (Neves et al. 2006).

[1]Bartneck et al. 2009 have argued that the Uncanny Valley hypothesis is not empirically valid. However, their own study is methodologically questionable and not statistically firm because it is centered on a single robot model, Geminoid HI-1. The subsequent design changes at HIL indicate that the Uncanny Valley hypothesis remains to be an important criterion for the developers.

Social robots seem especially well suited to aid in this task as their typical use cases demonstrate. They are used as therapeutic tools in the interaction with children with autistic spectrum disorders and elderly people suffering from dementia—two scenarios in which simplifications are essential to make communication possible in the first place (Dautenhahn 2003; Shamsuddin et al. 2012; Diehl et al. 2012; Broadbent et al. 2009). Here, an interesting diversion from the more common trend to build anthropomorphic robots can be observed. Generally, anthropomorphic appearances, behaviors, and modes of interaction are greeted in the design of social robots as very effective facilitators of social interaction (Fink 2012). People tend to react more positively toward robots exhibiting humanlike facial expressions and emotions as opposed to purely functional designs (ibid., p. 201), which is why the utilization of anthropomorphic features has usually been embraced by the research community within certain limits—including the Uncanny Valley problem and the problem of raising too high expectations on the side of human interactants (Duffy 2003). This finding has often been explained with people's tendency to make sense of unfamiliar phenomena by projecting humanlike characteristics onto them in order to rationalize their behavior (ibid., p. 180). In the case of people with autistic spectrum disorders or dementia, however, the biggest problem seems to consist in a diminished ability to correctly interpret complex human behavior. This is why design considerations in these fields differ from the more common approach of anthropomorphization in human–robot interaction.

Comparable to animal therapy, social robots can be employed to reduce social exchange to the very basics by filtering out all the noise and clutter that usually occurs in interpersonal contact. Social robots have the advantage of unambiguousness as has been very convincingly shown by an experiment that compared the levels of engagement of children with autism with different kinds of 'robots': a small humanoid doll in a humanlike and a plain variant and a professional mime artist, dressed up either as a robot or as a regular person (Robins et al. 2006). The children preferred the robotic variant in both cases, supporting the hypothesis that reduction of complexity is indeed the core mechanism that makes robots attractive to autistic children in the first place.

> Contrary to the highly complex and potentially unpredictable behavior exhibited in peoples' social interactions, the use of robots in the Aurora project allows for a simplified, safe, predictable, and reliable environment where the complexity of interaction can be controlled and gradually increased (ibid, p. 482).[2]

The authors of the study express as their future "aim to incrementally guide the children with autism to interact with robots toward more complex and human-like behavior" (ibid., p. 506).

Another widely used technique in social robotics research is the generation of facial expressions in robots following standardized psychological catalogues like the Facial Action Coding System (Ekman and Friesen 1978). Here, too, a drastic

[2]The results of the study influenced the design of the KASPAR robot developed at University of Hertfordshire.

reduction of complexity is achieved by limiting the range of human emotional expressions to a set of archetypes that is easily implementable and recognizable. For example, Tsuruda et al. (2013) have proposed a supplement to the Elfoid device—a mobile phone-sized telepresence android developed at HIL—in order to project cartoon emotions like anger, happiness, sadness, fear, etc., onto the robot's face. The basic technique here is that of caricature—by exaggeration and over-simplification certain traits are stressed and others neglected. It is to be expected that such simplified conversational settings work well in interactions with autistic children and/or demented people (cf. the conclusion to the literature review in Diehl et al. 2012). But research in social robotics is by no means restricted to these areas. The next section of this chapter reflects on the possible feedback effects that might result if interactions with robots become a regular element of social exchange—e.g., as imagined in Cynthia Breazeal's (2002) vision of sociable robots that has been quite influential in the research community.

7.3 Standardization of Social Situations

One of the grand research goals in social robotics is to make human–robot interaction 'natural' and intuitive, responsive to dynamic contexts and socially sophisticated. "Social interaction is not just a scheduled exchange of content, it is a fluid dance between the participants" (Breazeal 2002, p. 173). Looking at actual interactions with robots at the current level of technical sophistication though, it is hard to avoid the impression that this is not what is happening. Contrary to the intended trajectory—social robots gradually becoming more socially intelligent—people interacting with the current generation of robots tend to adapt their own behavior to tune into the robot's limited expressive capabilities and internal states. What is described as sociality by robotics researchers often resembles a formal game-like structure with predefined categories that are usually transparent enough to allow people to formalize themselves into adequate interactional partners for the robot.[3] At best, we deal with a kind of aseptic sociality, a clinically controlled space of registered exchanges that leave little room for the emergence of social complexity. Accordingly, a recent survey found that a majority of the participating staff of a disability service organization do not see social robots as humanlike social actors at all, but rather compare them to other technical appliances with specific tasks like coffee machines or vacuum cleaners (Wolbring and Yumakulov 2014).

Interestingly though, a lot of people seem to actively look out for this type of minimally social interactions. Social psychologist Sherry Turkle has described the transition to what she calls the "robotic moment" in society in great detail. In her analysis of people's interactions with artificial agents she discerns a specific

[3]Gillespie (2014) has made a similar argument referring to the process of algorithmic profiling, e.g. in content recommendation services and online shopping.

quality of interaction that she calls the ELIZA effect—"that desire to cover for a robot in order to make it seem more competent than it actually is" (Turkle 2011, p. 131). An 'as-if' mode of interaction prevails that leads to people engaging in a play-like encounter in "complicity with the machine" (ibid., p. 24). Turkle has also directed her attention to case studies dealing not only with social robots, but with online communication behavior and the management of social contacts via mobile devices, among others (Turkle 2008; Turkle et al. 2006). Here, she finds similar communicative patterns at work that are more akin to an elaborate simulation of sociality consisting of game elements and processes of administrative management. The great advantage of this type of interactions lies in their risk-free and highly flexible nature, indicating "a certain fatigue with the difficulties of dealing with people" (Turkle 2010, p. 5). A similar point has been made by Sparrow and Sparrow (2006) who additionally argue that the success of many contemporary social robots is based on a scheme of deception by design that we have an ethical imperative to avoid. Blackford (2012) has objected to this line of thinking by pointing to the usefulness of minor consensual illusions that may actually signify healthy social behavior. Anthropomorphizing robots may belong to this type of innocuous self-indulgence and willing suspension of disbelief.

Leaving aside these psychological and ethical considerations for the moment, one may ask for the model of communication underlying the reformatting of the social observed by Turkle. Drawing on cultural theory, the next section argues that parallel to the process of domestication of robots into the social world that we are currently witnessing, a different kind of domestication is taking place. In this perspective, social robotics can be seen as the engineering branch of a specific mode of political regulation.

7.4 The Cybernetic Hypothesis and Its Impact on Social Robotics Research

Again, Breazeal (2002) delivers the prompt: when introducing the need for sociable robots, she considers them as helpful tools in scientific inquiry to the extent that they can model human behavior. By comparing data generated in controlled robot experiments, insights into social behavior may be gained with the ultimate aim of regulating the social order.

> Furthermore, given a thorough understanding of the implementation, parameters of the model could be systematically varied to understand their effects on social behavior. By doing so, social behavior disorders could be better understood, which in turn could aid in the development of effective treatments (ibid., p. 1).

In this context, she refers to existing research on autism and the possible use of robots for therapy (Dautenhahn 2000). Autism is an especially interesting case for the argument made here as it is typically characterized by an unwillingness or inability to participate in communication and further social interaction.

Tiqqun (2001) have proposed a radical critique of a mode of governance they call the cybernetic hypothesis. They describe a paradigmatic shift in the way control is organized when the cybernetic hypothesis supplants liberalism. "Contrary to the latter, it proposes to conceive biological, physical, and social behaviors as something integrally programmed and reprogrammable" (ibid., p. 4). Historically, cybernetics was a field of interdisciplinary research formed around groups of scientists in post-WWII US military-sponsored programs. Some of the main premises of cybernetics—the study of "control and communication in the animal and the machine" (Wiener 2007)—are the structural homology of living and nonliving systems, a holistic approach to knowledge across disciplinary boundaries, and a transition from the observation of linear processes to circular causality and nonlinear dynamics, which would later be described as feedback mechanisms that lead to a self-regulation of system behavior. As an academic field, cybernetics has not survived into the present, with funding and institutionalization peaking in the 1960s and going down throughout the 1970s (Pias 2004). As an epistemology and consequently a political model of management, it is still thriving thanks to various exports from the sciences to the realm of social policy (e.g., Deutsch 1963; Beer 1967).

Sociocybernetics have subsequently become normalized and are today an integral—if mostly unconscious—rationale of political, economic, educational, judicial, and health care institutions. As Dieter Mersch has stated with reference to Tiqqun (2001), the societal function of cyberneticization lies in the creation of self-disciplined citizens (Mersch 2013, p 79f).

> Each person was to become *a fleshless envelope*, the best possible conductor of social communication, the locus of an infinite feedback loop which is made *to have no nodes*. The cyberneticization process thus completes the "process of civilization" to where bodies and their emotions are abstracted within the system of symbols (Tiqqun 2001, p. 18).

Frictionlessness and smoothness of the flows of information and communication become an imperative of the highest order in an information economy. It is through these means that sufficient transparency can be provided to sustain the self-regulatory mechanisms of the system. This is precisely the model of communication underlying social systems theory which has drawn extensively on works by authors working in the field of cybernetics, among them Heinz von Foerster, Humberto R. Maturana, and Francisco J. Varela (Paetau 2013, p. 86). In Luhmann's theory of society, communication is the basic element of social reproduction and every blockage of communication and/or obstacle in the creation of follow-up conversations hinders it. It is not so much the quality of communication that is decisive but its pure existence as a symbolic exchange. Reducing the complexity of informational flows via standardization and conventionalization is thus a prime mechanism of guaranteeing their perpetuation.

In a self-regulated society as imagined by sociocybernetics—and it is my argument that social robotics research is at the forefront of its realization—autistic spectrum disorders are simply unacceptable. Individuals absorbed in their own inner mental states act as information sinks, they do not process external information and do not feed it back into the flow. In their nonparticipation, they resemble the ancient figure of the *idiōtēs* (ἰδιώτης), the private man who does not take part in public affairs. As such, he is an element of social disorder, a potential source of noise and disruption, and has to be returned to a state of readability like a damaged hard drive. Whereas historically, they would have been locked up in a mental asylum or penitentiary, primarily with the goal of removing them from the surface of public discourse (Foucault 1965, 1979), the current mode of regulation proceeds differently. Taking into account the notion of reprogrammability, the right stimuli are determined to transform the suspected individuals into active participants of the fully transparent social order. The goal is to make them accessible again for the capture and flow of information and thus addressable for further communication.

Accordingly, social robots can be seen to fulfill a double function in a sociocybernetic regime. On an ideational level, social robots are the perfect citizens as they are infinitely reprogrammable and adjustable. Secondly, they serve as platforms for social experimentation and as a means of integrating those subjects whose self-regulation fails from a systemic perspective, whose observations elude second-order observations. As it is individual human beings with heterogeneous weaknesses who do not live up to systemic expectations, a society of—metaphorical or actual—robots can begin to seem like a desirable model. In this respect, "the cybernetic hypothesis is today the *most consequential anti-humanism*" (Tiqqun 2001, p. 6). The cybernetic hypothesis allows for "two kinds of scientific and social experiments":

- "to *turn living beings into machines*", and
- "to *imitate the living with machines*" (ibid, p. 13).

In its behavioristic focus (Ashby 1956), cybernetics has from its inception favored the analysis of forms of operation compared to questions of essence, nature, or materiality (Paetau 2013, p. 78). Cyberneticians, "biologists, doctors, computer scientists, neurologists, engineers, consultants, police, ad-men" share the common fantasy of a "*Universal Automaton*" (Tiqqun 2001, p. 13) that is a well-functioning element of the all-embracing feedback loops that make up society.

7.5 Conclusion

I have argued throughout the chapter that social robotics invests in specific types of complexity reduction to make possible human–robot interactions that resemble interpersonal encounters. In many instances, some of which have been discussed in the preceding pages, robots are built that are humanlike in a very rudimentary way so that people interacting with them are able to project additional traits as a way of personalization. Social complexity in human–robot interactions can be controlled to a very high degree which has some therapeutic value, for example in the treatment of autistic spectrum disorders or dementia. Whereas the vision of sociable robots aims at incrementally increasing the complexity of these robots to make them more acceptable for society and thus find new use cases, a different trajectory seems equally likely. As Turkle and others have observed, a standardization of social situations is taking place in robotics but also in other areas of mediated telecommunication. This raises urgent ethical and political questions, but from the standpoint of cultural theory, another aspect might be even more important. Social robotics research shares a number of premises with the cybernetic hypothesis as identified by Tiqqun and can be described as the engineering branch of a contemporary regulatory regime. As they facilitate the flow of information and (tele-)communication and make accessible sealed-off minds, social robots find their place as useful tools in the propagation of limitless communication and transparency of observations.

The field of social robotics is operating in very sensitive areas where the risk of dehumanization is always close at hand. It is thus recommendable that the research community pay close attention to the ways sociality is rendered in their experiments and how they can avoid to contribute to a total cyberneticization of all forms of interaction. The first and most important question in the field should always be directed at target groups' needs. There is a real danger of robot experiments being run with no clear benefit to the affected parties other than a stimulation of the robot-producing industry. Apart from such an obvious inversion of means and goal, one might consider how relational ties and affective bonds between people can be strengthened without resorting to robotic facilitators. If the problem-solving approach is narrowed down to the question of what type of robot would be best in any given situation, other approaches may not even be considered at all. Additionally, having otherwise passive people interact with robots might seem like a therapeutic success, but more research is needed to ensure that these interactions are actually according with participants' individual needs. Calling robots 'social' does not ensure that their behavior is always interpreted as such. When social interaction becomes just another mechanical task to accomplish for residents of a care facility, it might not bring about the positive effects on mental well-being it is supposed to. Whereas the idea of the *Universal Automaton* can be seen as part of a unifying transhumanist endeavor, that is the creation of the perfect citizen, a better benchmark for the quality of social research might be the amount of diversity it is capable of handling. This may at times include the option of nonparticipation—to abstain from communication altogether.

References

Advanced Telecommunications Research Institute International (ATR) (2012) Hugvie: project site. http://www.geminoid.jp/projects/CREST/hugvie.html. Accessed 14 March 2015

Ashby WR (1956) An introduction to cybernetics. Chapman and Hall, London

Bartneck C, Kanda T, Ishiguro H, Hagita N (2009) My robotic doppelgänger—a critical look at the uncanny valley. In: Proceedings of the 18th IEEE international symposium on robot and human interactive communication. Piscataway, NJ, pp 269–276

Beer S (1967) [1959] Cybernetics and management, 2nd edn. English Universities Press, London

Blackford R (2012) Robots and reality: a reply to Robert Sparrow. Ethics Inf Technol 14(1): 41–51. doi:10.1007/s10676-011-9266-6

Breazeal CL (2002) Designing sociable robots. Intelligent robots and autonomous agents. MIT Press, Cambridge

Broadbent E, Stafford R, MacDonald B (2009) Acceptance of healthcare robots for the older population: review and future directions. Int J Social Robot 1(4):319–330. doi:10.1007/s12369-009-0030-6

Dautenhahn K (2000) Design issues on interactive environments for children with autism. In: Proceedings of the third international conference on disability, virtual reality and associated technologies (ICDVRAT 2000). The University of Reading, Reading, pp 153–161

Dautenhahn K (2003) Roles and functions of robots in human society: implications from research in autism therapy. Robotica 21(4):443–452. doi:10.1017/S0263574703004922

Dautenhahn K, Nehaniv CL, Walters ML, Robins B, Kose-Bagci H, Mirza NA, Blow M (2009) KASPAR—a minimally expressive humanoid robot for human–robot interaction research. Appl Bion Biomech 6(3–4):369–397

Deutsch KW (1963) The nerves of government. Models of political communication and control, 2nd edn. The Free Press of Glencoe, London

Diehl JJ, Schmitt ML, Villano M, Crowell CR (2012) The clinical use of robots for individuals with autism spectrum disorders: a critical review. Res Autism Spectr Disord 6(1):249–262. doi:10.1016/j.rasd.2011.05.006

Duffy BR (2003) Anthropomorphism and the social robot. Robot Auton Syst 42(3–4):177–190. doi:10.1016/S0921-8890(02)00374-3

Ekman P, Friesen W (1978) Facial action coding system. A technique for the measurement of facial movement. Consulting Psychologists Press, Palo Alto

Fink J (2012) Anthropomorphism and human likeness in the design of robots and human-robot interaction. In: Ge SS, Khatib O, Cabibihan JJ, Simmons R, Williams M (eds) Social robotics: 4th international conference, ICSR 2012, Chengdu, China, 29–31 Oct 2012, Proceedings. Springer, Heidelberg, Berlin, pp 199–208

Foucault M (1965) Madness and civilization. A history of insanity in the age of reason. Pantheon Books, New York

Foucault M (1979) Discipline and punish. Vintage Books, New York

Gillespie T (2014) The relevance of algorithms. In: Gillespie T, Boczkowski P, Foot K (eds) Media technologies: essays on communication, materiality, and society. MIT Press, Cambridge, pp 167–194

JST CREST (2010) Research promotion program creation of human-harmonized information technology for convivial society. http://www.geminoid.jp/projects/CREST/index.html. 14 March 2015

Knodt EM (1995) Foreword. In: Luhmann N Social systems. Trans. Bednarz J, XI–XXXVI. Stanford University Press, Stanford

Mersch D (2013) Ordo ab chao—order from noise. Diaphanes, Zürich

Mori M (1970) The uncanny valley. Energy 7(4):33–35

Neves C, Eckert B, Neves FM (2006) What is complex in the complex world? Niklas Luhmann and the theory of social systems. Trans Marcelo Severo Sociol 15:182–207

Paetau M (2013) Niklas Luhmann and cybernetics. J Sociocybern 11(1–2):75–103

Pias C (2004) Zeit der Kybernetik—Eine Einstimmung. In: Pias C (ed) CYBERNETICS. The Macy-conferences 1946–1953. Essays and documents, vol 2. Diaphanes, Berlin, pp 9–41

Robins B, Dautenhahn K, Dubowski J (2006) Does appearance matter in the interaction of children with autism with a humanoid robot? Interact Stud 7(3):479–512

Shamsuddin S, Hanafiah Y, Luthffi II, Salina M, Fazah AH, Nur IZ (2012) Initial response in HRI—a case study on evaluation of child with autism spectrum disorders interacting with a humanoid robot NAO. Procedia Eng 41:1448–1455. doi:10.1016/j.proeng.2012.07.334

Sparrow R, Sparrow L (2006) In the hands of machines? The future of aged care. Mind Mach 16(2):141–161. doi:10.1007/s11023-006-9030-6

Tiqqun (2001) L'hypothèse cybernétique. *tiqqun* (2). English edition: the cybernetic hypothesis. http://www.mediafire.com/view/2i1vkj39l25i4c2/cybernet.pdf. Accessed 14 March 2015

Turkle S (2008) Always-on/always-on-you: the tethered self. In: Katz JE (ed) Handbook of mobile communication studies. MIT Press, Cambridge, pp 121–137

Turkle S (2010) In good company? On the threshold of robotic companions. In: Wilks Y (ed) Close engagements with artificial companions: key social, psychological, ethical and design issues. John Benjamins, Philadelphia, pp 3–10

Turkle S (2011) Alone together. Why we expect more from technology and less from each other. Basic Books, New York

Turkle S, Taggart W, Kidd CD, Dasté O (2006) Relational artifacts with children and elders: the complexities of cybercompanionship. Connect Sci 18(4):347–361. doi:10.1080/09540090600868912

Tsuruda Y, Hori M, Yoshimura H, Iwai Y (2013) Generation of facial expression emphasized with cartoon techniques using a cellular-phone-type teleoperated robot with a mobile projector. In: Human-computer interaction. Towards intelligent and implicit interaction: 15th international conference, HCI international 2013, proceedings, Part V. Springer, Berlin, Heidelberg, pp 391–400

Wada K, Shibata T (2006) Living with seal robots in a care house: evaluations of social and physiological influences. In: IROS 2006: IEEE IRS/RSJ international conference on intelligent robots and systems. Piscataway, NJ, pp 4940–4945

Wiener N (2007) [1948] Cybernetics: or control and communication in the animal and the machine, 2nd edn. MIT Press, Cambridge

Wolbring G, Yumakulov S (2014) Social robots: views of staff of a disability service organization. Int J Soc Robot 6(3):457–468. doi:10.1007/s12369-014-0229-z

Chapter 8
Open Sourcing Social Robotics: Humanoid Artifacts from the Viewpoint of Designers

Davide Fornari and Serena Cangiano

The aim of the paper is to define a remit of interest for design and designers in the field of products and services employing human face as an interface with users. From the point of view of design culture, humanoid artifacts, are at the intersection of product design, visual communication, and interaction design. The release of open hardware platforms, such as Arduino Robot, as well as the diffusion of fabrication laboratories (fablabs) and open design practices are the early manifestations of a wider interest toward social robotics accompanied by the possibility of building working prototypes of humanoid artifacts. The use of human face as interface in human–computer interaction adds analogic content to interaction processes in an unmediated and unobtrusive way, transferring human–computer interaction to the field of interpersonal communication. Nevertheless, the field of anthropomorphic artifacts is generally neglected by designers and rather controlled by engineers and experimental psychologists. Starting from a definition and taxonomy of humanoid artifacts, the aim of this chapter is to describe the paradigm of human–computer interaction through case studies of humanoid artifacts and to define criteria for the evaluation of humanoid artifacts from the perspective of design practice.

D. Fornari · S. Cangiano (✉)
Laboratory of visual culture, SUPSI University of Applied Sciences and Arts of Southern Switzerland, Canobbio 6952, Switzerland
e-mail: serena.cangiano@supsi.ch

D. Fornari
e-mail: davide.fornari@supsi.ch

J. Vincent et al. (eds.), *Social Robots from a Human Perspective*,
DOI 10.1007/978-3-319-15672-9_8

8.1 Introduction

Tomás Maldonado considers the proliferation of automata during the eighteenth century as a form of struggle against the dualism between practical and theoretical knowledge. At this stage a debate on the utopic role of machines and on their capacity of improving people's lives began: the association of technical objects to human morphology is envisioned as a way to "favour the trend of considering machines as a model for human beings" (Maldonado 2005).

The main effort for scientists, artists, and craftsmen dealing with humanoid artifacts, which in the past meant mainly automata, was to make the idea of an anthropomorphic machine socially acceptable in order to eventually host it within the domestic landscape. This technological translation of the human body, which is at the origin of modern robotics in Japan, had to face for long the opposition of the Catholic church (Maldonado 2003). This was connected with pagan and Jewish imaginarium of beings created by Man, such as Golem, that can be defined as 'dreamlike mechanics,' which are evidently nonfeasible but of high impact on culture (Hornyak 2006; Maeda 2005). The path toward social acceptance of anthropomorphic machines and objects can be read across time, and it shows how religions and ethics had an influence over the status of technical objects and the community connected with their production (Hornyak 2006; Legrenzi 2002).

The aim of this chapter is to define key theoretical issues regarding the field of robotics from the viewpoint of designers. The increasing availability of open-source robot hardware allows the participation of large audiences in the design of social robots and humanoid artifacts. By 'humanoid artifacts' we refer to serial products or products serially used (i.e., diffused or used through information and communication technologies (ICTs)), which incorporate technology or offer technological performances and employ a facial configuration as an interface.

8.2 A Taxonomy of Humanoid Artifacts

A classification of humanoid artifacts is required to offer sound tools for the evaluation of human–machine interaction. These artifacts entail issues connected with human brain cognitive specificity in the perception and recognition of face, as well as psychology of communication. The perceptual misunderstanding of face recognition without the presence of a real human face, but in front of a reproduction of face, is the advantage of humanoid artifacts: the face offers the feeling of presence to users.

If we accept anthropomorphic configuration of face in humanoid artifacts as a specific feature, it is important to evaluate in which typologies of artifacts it is employed. For this reason, we propose a taxonomy of humanoid artifacts, working both for historical and currently produced devices, based on dimensions (2D or 3D) and movements (still or moving) (see Fig. 8.1). The reproduction of face has specific features connected with the physical or perceptual support: discrimination between static and dynamic artifacts, as well as bidimensional and three-dimensional ones, is of main concern.

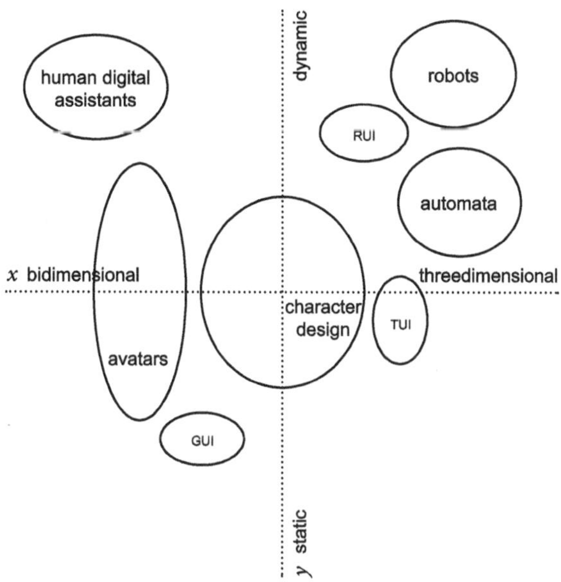

Fig. 8.1 Taxonomy of humanoid artifacts: on *x*-axis dimensions, on *y*-axis movement. GUI stands for Graphical User Interface, TUI for Tangible User Interface, RUI for Robotic User

Technologies employed in the reproduction of face influence the efficiency of humanoid artifacts. Realistic avatars used in social networking software such as Second Life are rendered graphical objects (bidimensional, yet looking three-dimensional) that risk losing their efficiency when movements are slow as a result of low performance (RAM, data bandwidth) of the hardware hosting the interaction.

Realistic androids freely moving were impossible to build so far, due to the quantity of compressors needed to manage the actuators for face expressions: they are limited to a semi-sitting position. Yet, the diffusion of hardware for 3D prototyping and PVC stamping are making the production of three-dimensional chassis a relatively low investment, and they allow a high quality, even for limited print runs.

8.3 The Impact of the Open-Source Movement on Building Robots

Open hardware and open design have made available physical and technological artifacts whose information and implementation specifications (blueprints, code, material bills) are released to the public under free licenses. These latest developments, together with digital production techniques and the increasing "maker movement" (born in the early 2000s), are triggering the diffusion of new practices of design and production of social robots.

These practices are based on the possibility for designers and experts of technology, as well as for do-it-yourself amateurs, to interact with on-line communities, platforms, and collaborative services, with the aim to use and/or share digital files (hardware schemes, software codes, 2D and 3D files) of a robot, and to allow for modification and production through technologies and low-cost rapid prototyping services. The practices connected with open source are the epiphenomenon of the passage from postmodernity to a participative era (typical of peer to peer approach) where the culture of "do it together" is made possible by ICT and in the specific case of physical artifacts, such as social robots, by production technologies.

The development of open-source hardware became a viable solution in the year 2000 thanks to three factors (Cangiano and Romano 2015): the availability of larger bandwidth for accessing the Internet and exchanging large source files for hardware projects; the reduction of the costs for hardware technologies and their production; and the commercial success of open hardware projects such as Arduino, Little Bits, RepRap, and Sparkfun that conveyed to a larger public the benefits of such a business model.

The key features of the development of open-source hardware can also be condensed into three (Gibb 2014): open-source hardware as an infrastructure for the *production of devices* not yet available on the market (innovation enabler); open-source hardware as an infrastructure for *empowering people's communities* in the collaborative solution of complex problems and socially relevant problems (open-source hardware as user and community empowerment tool); and open-source hardware as an infrastructure *determining the simplification* of technology (technology mediator).

The protagonists of open-source hardware have shown a clear interest in robots. The company Arduino, producer of the open-source electronics platform based on easy-to-use hardware and software, released in May 2013 the Arduino Robot, a programmable device with a couple-wheeled board mounting two processors, one controlling the motors and the other reading sensors and controlling actuators. Differently from traditional firms that develop and design robots, the open hardware companies such as Arduino propose a new approach that considers robots as a means for people to have a sense of ownership of the technology whilst figuring out a technological future that is close to nonexpert people. For example, the open hardware famous claim "if you cannot open it, you do not own it" is the basic concept that was spread within the maker community for communicating a new way of interacting with devices in everyday life. Built upon this concept, Brian David Johnson, Intel's resident futurist, reflected on what makes a twenty-first century robot different from a twentieth century robot arriving at the notion that: "I'd say that a twenty first Century Robot is imagined first. It's certainly social, open source, and iterative, but also filled with the hopes and dreams of the people that made her, him or it. It's also easy to build!" (Johnson 2014).

8.4 Theoretical Issues: An Agenda for Designers Approaching Robotics

Starting from the definition and taxonomy of humanoid artifacts, we intend to highlight a number of crucial, theoretical issues that designers should evaluate and assess while approaching the design and production of social robots, due to the increasing availability and distribution of open technologies and designs.

Intuitiveness The simplicity and intuitiveness of an artifact depends on how easily the user can benefit from it with little or without any prior instruction. This implies the use of metaphors familiar to users, such as models for use easily added to the principal function of the device or service, or communication and interface systems which are uncomplicated or based on innate metaphors such as in analog and nonverbal communication. Simple interfaces and devices are the ultimate aim of every design to reduce functional complexity and to guarantee an acceptable degree of usability to the nonexpert end user. Often entire product families have seen a reduction of signs in the interfaces, reversely proportional to the increase in the functional complexity offered by devices, such as in the Apple iPod (Maeda 2005). Nowadays, the use of interfaces and simple interaction systems based on innate models is widespread in every field of application. The intuitiveness and simplification of devices has promoted easy, untrained, nonacculturate forms of interaction, which have virtually eliminated the need for localizing product interfaces, thus contributing to the globalization in marketing products.

Pragmatic of Human Communication Face and facial expressions participate in the transmission of messages, according to the pragmatic of human communication. In the case of humanoid artifacts, interfaces can move within a range from abstraction to figurativeness, since we can recognize faces and characters even in slightly defined configurations of faces.

In order to structure the pragmatic of human communication, Watzlawick et al. (1967) define a number of axioms. Their aim is to describe the process of communication as a calculus, in the most precise way, through meta-communication. The first axiom of pragmatic of human communication is 'One cannot not communicate': every behavior is a kind of communication to someone who is eager to give an interpretation of this behavior. Both verbal (natural languages) and nonverbal communication (posture, gestures, and especially facial expressions) are constant source of messages. When a face is 'artificial', as a result of an action of design, it adds, mixes, and overlaps symbolic, communicational, and semiotic issues. The face can be considered as a stage where the individual parts (i.e., eyes, mouth, ears, nose, eyebrows, etc.) perform as actors in a theater company.

Analog communication encompasses nonverbal communication and symptom strategies (e.g., silence, sleep), thus including the role of face and its expressions in communicating the relationship aspect of a message. According to Watzlawick, the combined use of both digital and analog communication is specific to *homo sapiens* (fourth axiom). Humanoid artifacts enrich the dynamics of human–machine interaction by employing an innate metaphor: the analog communication of facial expressions.

Analog communication is envisioned as a more precise and accurate way of communicating in comparison with natural language, as it can express faster than language emotions such as delight, surprise, dislike (Maeda 2005). Nevertheless, analog communication presents advantages and disadvantages when employed in human–machine interaction: while the ability of recognizing a face and its expressions is innate, many 'symbols' of analog and nonverbal communication vary between cultures (Legrenzi 2002).

Facial Prominence In visual communication, images have a main role for their anaphoric nature: they repeat the object they depict through representation instead of language. Concerning the face, experimental psychology proved how the same face produces more socially desirable information when it is prominent compared to the rest of the body (Archer et al. 1983). The face-ism index was introduced in order to measure facial prominence on print and TV media. It is the ratio of two linear measurements: the distance between the top of the head and the chin (x), and the distance from the top of the head to the lowest visible part of the subject (y). The face-ism index formula is $x{:}y$ and its result can vary in a range between 0 and 1. Close-ups have a face-ism index of 1, while medium to full shots have decreasing values. A shot excluding the face has a face-ism index of 0.

In an experiment, people were asked to assess personal features, ability, and competencies of persons portrayed in the stimuli with different degrees of face-ism. As a result, the people represented with higher facial prominence were considered, despite gender, more intelligent, assertive, and attractive, compared with those with a lower face-ism index. The important research result on facial prominence is the positive social expectation of face images with high face-ism index (Costa and Ricci Bitti 2000). If applied to humanoid artifacts, these experimental results suggest that face representation drives the perception of users concerning secondary qualities, those depending on the interpretation of a subject, as defined by John Locke.

The Uncanny Valley The 'uncanny valley' is a diagram connecting familiarity and human likeness of humanoid artifacts. The hypothesis by Mori (1970), researcher and roboticist, was that the more a humanoid artifact resembles its model, the more likely users notice the small differences. Later scholars have largely equated this idea of human resemblance with the social acceptance of robots (e.g., Höflich and El Bayed in this book). In case of high resemblance, a response of revulsion is possible. The more a humanoid artifact is similar to its model, the more our sense of familiarity increases. This increase is higher in case of moving artifacts. Small but detectable differences put a stop to this increase, since artifacts are no longer perceived as familiar and pleasurable, but rather unfamiliar, nonpleasurable, eerie, and uncanny. Mori chose as an example a prosthetic hand, which can be a perfect replica concerning color, texture, mobility, but still is uncanny during a haptic interaction, since it is going to be cold at touch.

Mori's hypothesis was for long considered nonscientific, mainly because it was a theory based on partial artifacts (i.e., a prosthetic hand). According to Hiroshi Ishiguro, if a robot is very 'robotic,' we do not compare it with a human model in order to recognize it. But if a robot looks like a human being, then we recognize its model, and we notice the small differences between android and human being (Hornyak 2006). MacDorman (2005) established a connection between the uncanny features of technical objects and the fear of death in human beings: machines resembling their model too much unveil their artificial nature which can eventually be eerie, unexplainable, and scary for nonexperienced users (Bartneck et al. 2007).

Experiments on the uncanny features of highly resembling humanoid artifacts are based on the categorization of images by users. Seyama and Nagayama

(2007) employed morphing between real images and drawings in order to show how the uncanny valley hypothesis is confirmed for highly resembling artifacts in presence of abnormal features. Research conducted by Bartneck et al. (2007) highlights how the aspect of artifacts influences their resemblance to human models. They propose a correction to the uncanny valley diagram: the uncanny cliff, since there are a limited number of highly human-like robots. "It appears unwise to attempt to build highly human-like androids, since they would not be liked as much as more machine-like robots" (Bartneck et al. 2007, p. 372). But beyond the problem of 'familiarity' versus 'likeability', the issue is the concept of uncanny, which refers to eeriness. Highly human-like artifacts with a detectable artificial nature result as uncanny: if a machine's interface is eerie and unacceptable, no interaction is possible.

8.5 Case Studies

We adopted the instrumental case study methodology in order to examine design aspects connected with open-source humanoid artifacts. The selection of case studies is provided here for presenting and assessing different strategies adopted in open-source robot projects employing human face in human–machine interaction processes.

Arduino Robot Arduino robot[1] is a do-it-yourself and open-source platform for learning robotics and building robots (see Fig. 8.2). The robot is completely reprogrammable, thanks to a USB port that allows programs developed via Arduino IDE (integrated programming environment), the easy to use electronics platform, to be uploaded. It consists of two rounded printed circuit boards: the top board is the control and interface unit, the bottom board is the sensor unit featuring wheels. It has been designed to support the easy configuration of robotics applications so it features a number of integrated inputs, such as potentiometers, buttons, and floor sensors.

The users can program the behaviors of the robot by connecting a USB cable. Arduino Robot features sensors and actuators mounted on two printed circuit boards. The robot has two processors, one on each of its two boards. The microcontroller boards are based on the ATmega32u4. The users can add additional parts according to the project they intend to develop (e.g., eyes, arms, etc.) (McComb 2013; Margolis 2013).

3D Printed Animatronic Robot Head The Animatronic Robot Head is an electronic kit designed to provide people with a tool for learning about the 3D printing processes behind the production of the parts of a robotic body (see Fig. 8.3). It consists of a robotic puppet made of speakers and servo motors. The robotic head is controlled through the use of servo motors, easy to use and accessible electronic

[1]Arduino Robot, http://arduino.cc/en/Main/Robot (retrieved on 20/11/2014).

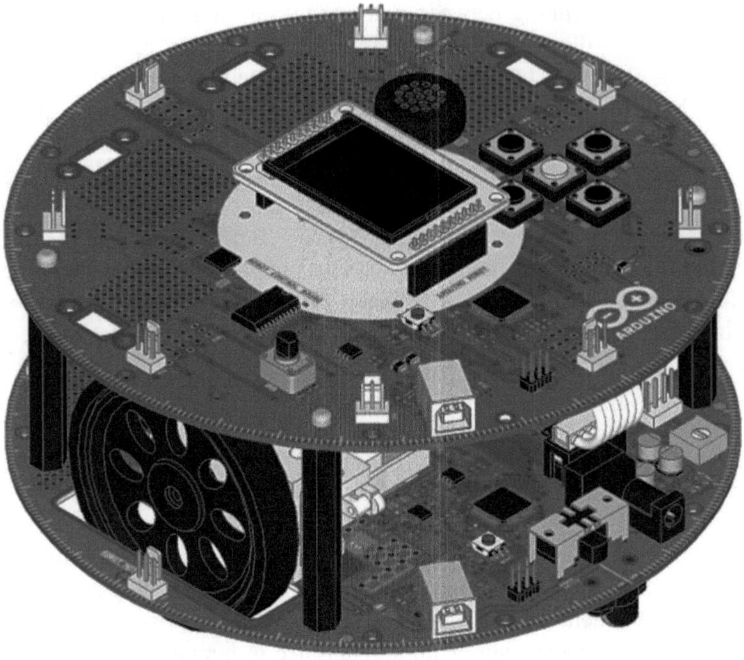

Fig. 8.2 Arduino robot, 2014

motors that are commonly available in the market for modeling hobbyists. The two servos are programmed for controlling the movement of the head while two speakers are used to make the robotic eyes and an LED mouth for a friendly remote-controlled robot. The parts are designed with Tinkercad,[2] an easy to use Web-based 3D modeling software, originally developed for children in order to lower the barrier for nonexperts.

iCub The iCub is an open-source cognitive humanoid robotic platform developed at IIT (Istituto Italiano di Tecnologia) as part of the EU project RobotCub ended in 2010, whose main goal was to study cognition through the implementation of a humanoid robot with the size of a 3.5 year old child[3] (see Fig. 8.4). iCub has 53 motors that move the head, arms and hands, waist, and legs. It can see and hear; it has the sense of proprioception (body configuration) and movement (using accelerometers and gyroscopes). The humanoid robot iCub is able to make human-like eye and head movements. Moreover, it can recognize faces and grasp

[2]*3D Printed Animatronic Robot Head*, https://learn.adafruit.com/3d-printed-animatronic-robot-head?view=all#mission-control (retrieved on 20/11/2014).

[3]*An Open Source Cognitive Humanoid Robotic Platform*, http://www.icub.org (retrieved on 20/11/2014); *An Open Framework for Research in Embodied Cognition*, http://www.robotcub.org (retrieved on 20/11/2014).

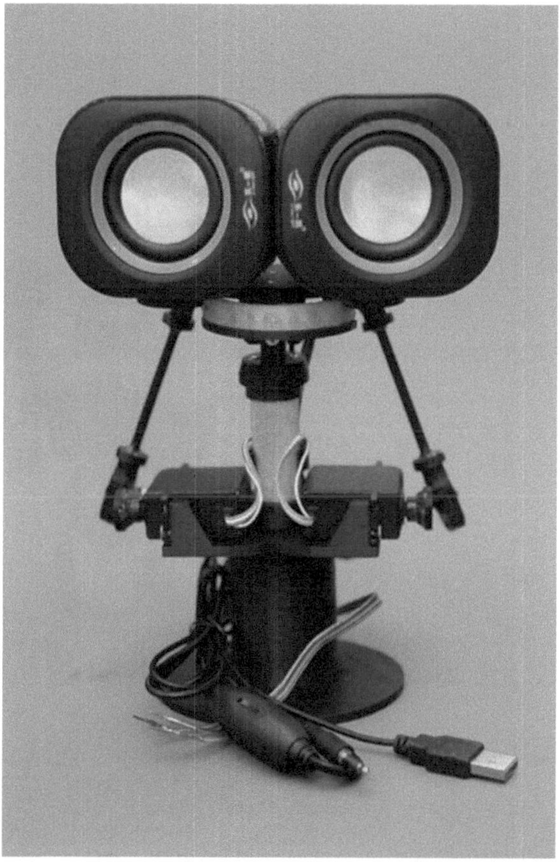

Fig. 8.3 3D printed animatronic robot head: an entry-level robotics/animatronics project by Adafruit, 2013–2014

Fig. 8.4 iCub robot during an affordances demo, 2010

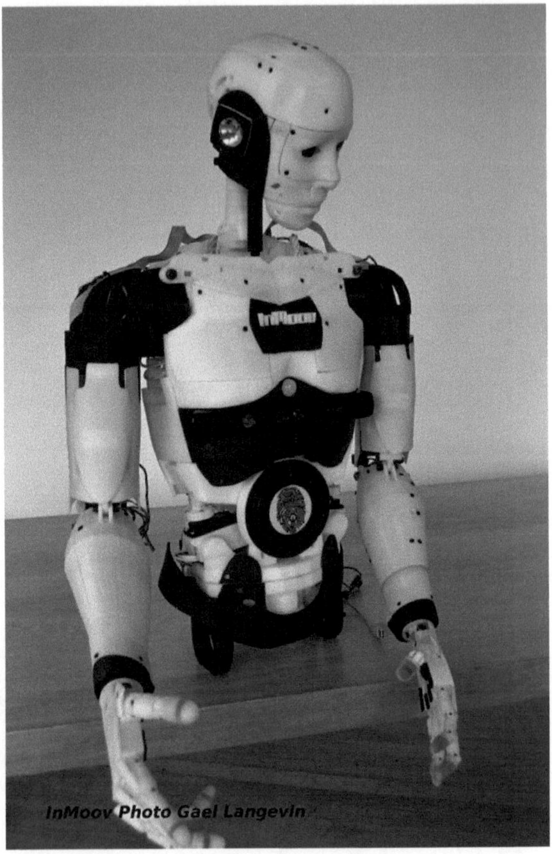

Fig. 8.5 Inmoov robot, 2012–2014

objects. The iCub project is open source: the platform is distributed openly and it features open-source software that can be downloaded for free from the website, modified, and improved. Thanks to its level of openness, the project has been subsequently adopted by more than twenty laboratories worldwide that develop research in robotics or robotics applications. Recently, the iCub has been further developed with the integration of force control, a skill that allows the robot to interact with people in a safe and gentle way. In order to minimize weight and cost, some of the iCub's parts such as the hand have been produced from acrylonitrile butadiene styrene (ABS) using 3D printing techniques. The choice of designing the parts for the additive manufacturing technology and processes removed the need for extensive machining which would add significantly to the overall cost of the robot (Tsagarakis et al. 2007).

InMoov InMoove[4] (see Fig. 8.5) is a life-sized android completely made out of 3D printed parts that can be downloaded from the most popular collaborative

[4]*Open-Source 3D Printed Life-Size Robot*, http://www.inmoov.fr (retrieved on 20/11/2014).

Fig. 8.6 Ono robot, 2012

3D model file sharing platform on the Internet, Thingiverse.[5] The creator of the project is Gael Langevin, a French sculptor and model marker who has spent years designing and engineering the animatronic robot. The InMoov project started with the design and development of an arm and a hand and it has reached half a body (Abid et al. 2014). The cost is around 900 US dollars. It runs on an Arduino, and all the parts can be produced with a DIY low-cost 3D printer with a 12 × 12 × 12 cm printing volume.

Ono Ono (Fig. 8.6) is an open-source social robot that can be produced using DIY tools and techniques[6] (Vandevelde et al. 2014). The actual robot consists of four main parts: the modules for the control of the facial features (eyes, eyebrows, mouth); the frame of the robot, made from laser-cut interlocking cross-sections that accommodate the sensors and actuators of the facial modules; foam and textile cover that create the software body of the robot and protects the inner part; the control unit and electronics that consist of a separate control box with joystick interface and a microcontroller that reads the joystick position in order to send the joystick values to the servo controller inside the robot. The robot has been tested with autistic children. One of the main goals of the project is to avoid the major obstacles in the study of human–robot interaction with social robots because there is the lack of platforms to allow for tests with large user groups. Since Ono robot's parts are soft, made with cheap materials and fabrics, it suits the context where the users have to touch or hug it.

---

[5]http://www.thingiverse.com/hairygael/designs (retrieved 21/2/2015).

[6]*Edubots: Ono*, http://io.workspace.howest.be/edubots/ono/ (retrieved on 20/11/2014).

8.6 Discussion and Conclusions

This chapter presents some key theoretical issues and a collection of case studies in order to define an agenda for the designers of humanoid artifacts. Increasing availability of open-source robot technologies and robot designs, and the rise of robot makers' movement (Baichtal 2014) created the need for this agenda setting. The work is based on the hypothesis that is due to specific features of face recognition processes, we tend to treat humanoid artifacts as characters, transforming human–machine interaction into something more similar to a process of communication between peers, hence falling into the pragmatics of human communication, and, further, that designers should have a larger impact and awareness in the domain of robotics.

The Arduino robot has made available to the largest community ever a platform for building self-moving robots, thus simplifying the experimentation on robotics and empowering designers. In general, the diffusion of open-source "kits" such as the 3D printed animatronic robot head, shall be saluted as a significant contribution to the diffusion of robots and robotic technologies, as it allows for robotics to enter an increasingly public discourse. iCub has been equally important in providing the experimental platform of a highly accomplished robot for groups of scientists other than its creators. InMoov can be acknowledged for its availability as a "downloadable" item that can be 3D printed in an environment that is increasingly familiar to that of professional designers: fablab. The project Ono was introduced to show how the prominence of human face and face actors can simplify the construction of a social robot, helping designers to focus on specific facial features rather than on a global similarity with human models.

Employing a multidisciplinary approach and gathering concepts from the fields of design, technology, history of applied arts, experimental psychology, and robotics, a number of qualitative and quantitative parameters can be offered for setting an agenda for designers and amateurs who want to tackle the issue of social robotics in their practice. The highlighted factors for assessing humanoid artifacts include:

- discussing the level of intuitiveness of an artifact;
- describing the interaction channels of the artifact from the viewpoint of the pragmatic of human communication;
- measuring the face-ism index of the artifact;
- defining the position of the artifact in the uncanny valley diagram.

These factors are far from being definitive, but they present one response to the lack of precise framework for assessment and design that is obvious in the rapidly developing and innovative field of robotics and human digital assistants. Compared with the example of mobile ICTs, the design of a robotic interface is far more complex. This complexity is both due to the lack of standards for robotics at this relatively early stage of development, and to the technological complexity of three-dimensional interfaces, mixing sensors, and actuators versus the bidimensional design of mobile device interfaces. If intuitiveness and familiarity must be attributes of interfaces in both cases (robots and mobile devices), the results are far more diverse in the case of robots due to the abovementioned lack of standards.

Nevertheless, open-source robots have a high potential in making social robots more acceptable for a variety of reasons: open-source technologies are less expensive and collaborative, thus multiplying the experimentations and diffusing the production and actual presence of robots out of the laboratory environment. Moreover, in open-source projects design, production and usability testing are merged as part of a design-oriented process, more focused on deployment of already existing technical solutions than on innovative technologies. The simultaneous design, production, and usability are also connected with a shared and communitarian approach, where not only designers test their products, but they take advantage of open on-line and off-line networks of scholars, designers, and amateurs. In this sense, the use of open-source technology and free licenses affects the very design process of robotics.

References

Abid MR, Meszaros Ph E, da Silva RF, Petriu EM (2014) Dynamic hand gesture recognition for human-robot and inter-robot communication. In: Proceeding of the 2014 IEEE international conference on computational intelligence and virtual environments for measurement systems and applications (CIVEMSA), May 5–7, 2014, pp 12–17

Archer D, Iritani B, Kimes DD, Barrios M (1983) Face-ism: five studies of sex difference. J Pers Soc Psychol 45:725–735

Baichtal J (2014) Robot builder: the beginner's guide to building robots. Que, Indianapolis

Bartneck C, Kanda T, Ishiguro H, Hagita N (2007) Is the uncanny valley an uncanny cliff? In: Proceedings of the 16th IEEE international symposium on robot and human interactive communication, IEEE, Jeju, 2007, pp 368–373

Cangiano S, Romano Z (2015) Open sourcing wearables: the impact of open technologies and user engagement in the design of body-borne interactive products. In: Bihanic D (ed) Empowering users through design. Interdisciplinary studies and combined approaches for technological products and services. Springer, London

Costa M, Ricci Bitti PE (2000) Face-ism effect and head canting in one's own and others photographs. Eur Psychol 4:293–301

Gibb A (2014) Building open source hardware: DIY manufacturing for hackers and makers. Addison-Wesley Professional, Boston

Hornyak TN (2006) Loving the machine: The art and science of Japanese robots. Kodansha, Tokyo

Johnson BD (2014) 21st century robots. Maker Media, Sebastopol

Legrenzi P (2002) La mente. Il Mulino, Bologna

Maeda J (2005) The laws of simplicity. MIT Press, Cambridge

MacDorman KF (2005) Androids as experimental apparatus: why is there an uncanny valley and can we exploit it? In: CogSci 2005 Workshop: toward social mechanism of android, Stresa, pp 106–118

Maldonado T (2005) Disegno industriale: un riesame. Feltrinelli, Milan

Maldonado T (2003) The body: artificialization and transparency. In: Fortunati L, Katz JE, Riccini R (eds) Mediating the human body: technology, communication, and fashion. Erlbaum, Mahwah, pp 15–22
Margolis M (2013) Make and Arduino-controlled robot. O'Reilly, Sebastopol
McComb G (2013) Arduino robot bonanza. McGraw-Hill Professional, New York
Mori M (1970) The uncanny valley. Energy 7(4):33–35
Seyama J, Nagayama R (2007) The uncanny valley: effect of realism on the impression of artificial human faces. Presence 16(4):337–351
Tsagarakis NG, Metta G, Sandini G, Vernon D, Beira R, Becchi F, Righetti L, Santos-Victor J, Ijspeert AJ, Carrozza MC, Caldwell DG (2007) iCub: the design and realization of an open humanoid platform for cognitive and neuroscience research. Adv Robot 21(10):1151–1175
Vandevelde C et al. (2014) Ono, a DIY open source platform for social robotics. In: Proceedings of TEI 2014, 16–19 Feb 2014, Munich, Germany
Watzlawick P, Helmick Beavin J, Jackson DD (1967) Pragmatic of human communication. A study of interactional patterns, pathologies and paradoxes. Norton & Co, New York

Part III
Social Robots in Everyday Life

Chapter 9
The Mobile Phone: An Emotionalised Social Robot

Jane Vincent

This chapter builds on the notion that humans, who have appropriated mobile phones and incorporated them into their everyday lives since the 1980s have, in so doing, created their own personal social robot. It asserts that the constant always on connectivity afforded by this device is enabling a communicable stream of consciousness and emotions that are intertwined between the mobile phone and their emotional self. This, in turn has created a dependence and attachment to the device, to the relationships it mediates and more, such that it is so fully integrated into people's day-to-day living they cannot imagine how to conduct everyday life without it. The outcome of this human and machine interaction, and the electronic emotions it imbues, is a device that has become an emotionalised social robot that is exclusive to its user.

9.1 Introduction

In this chapter I explore the proposition that the mobile phone has become an emotionalised social robot and ask how has the human machine interaction involved in using a mobile phone transformed it thus. My previous research (Vincent 2013) has asserted that the personalised mobile phone and its user combine to co-create a social robot. In this chapter I develop my exploration of this unique bond and symbiotic relationship to examine the importance of emotion in this context. I will argue that the mobile phone is not only a personalised social robot but, furthermore, it is so emotionally driven and charged with the feelings of

J. Vincent (✉)
University of Surrey, Surrey, UK
e-mail: j.vincent@surrey.ac.uk

J. Vincent et al. (eds.), *Social Robots from a Human Perspective*,
DOI 10.1007/978-3-319-15672-9_9

its user that it is an emotionalised social robot. This is made possible by the flexibility of the mobile phone systems and applications that allow its user to shape and reshape its functionality to meet their personal needs. Touch, hold, gaze, use of headphones, vibration and physical mobility (co-located with the user) are just some of the haptic properties of the mobile phone that stimulate action and electronic emotions. This flexibility of design makes the mobile phone a more dynamic and adaptable machine than many of the more rigid and purpose built social robots such as AIBO, Paro or Hugvie as will be discussed later.

Contemporary literature by others on human robot interaction includes exploration of emotion from biological, cognitive and philosophical perspectives. For example Nitsch's and Popp's (2014) cognitive examination of the expression of emotions in robots and Ferrando's (2014) gendered post-human perspective have recently added new research to the debate on emotion and robot design. Ferrando's suggestions that contemporary research '*relocates the discourse within a symbiotic paradigm rather than a dualistic one*' and that '*robots are going to evolve in unique and peculiar ways which are hard to predict*' (Ferrando 2014) highlight the diversity of research and the opportunities for multidisciplinary developments in this field of study. Emotion associated with a social robot typically occurs when the emotion has been designed and engineered into the robot to emulate or respond to specifically identified human needs. However, in Syrdal et al's (Syrdal et al. 2007) study of a personalised social robot companion they report the familiarity that develops between a human user and a robot which affirms the role of emotion in establishing the companion status of the robot: '*individual differences change as participants become more familiar with the robot*' (p. 1147).

A further point to consider in this introduction is that the mobile phone as an emotionalised social robot is different from the robot designed and engineered to be personalised and empathetic to its human user (Dautenhahn 2004) in that the mobile phone has not been intentionally designed with robotic functionality. Rather it is a communications and entertainment device that has multiple functionalities including a miniaturised computer, telephone, television, games console, camera and video recorder which are appropriated and personalised in a unique way by its user.

In this chapter I explore the robotic turn within the human that is determined by the human user's intuitive emotional responses to everyday life activities transacted via their mobile phone. These emotional triggers form the basis of the communicative practices and mediated actions delivered via their mobile phone. I develop this argument further by suggesting that this relationship is dominated by the user's emotions. Furthermore, the reason for their personalisation practice is that it enables mediation of emotions between the user and the content on their mobile phone, as well as the communications they transact (in any human–human/machine–machine combination).

The discussion in this chapter is framed by the theoretical concepts of the presentation of self (Goffman 1959) and electronic emotions (Vincent and Fortunati 2009). The chapter draws on various research studies conducted since 2003 by the author (cf Vincent 2003, 2006, 2009, 2010, 2014). These studies examined this close attachment people have with their mobile phone and provide the basis for the assertion herein that the mobile phone has become an emotionalised social robot. The background to the development of both mobile phones and social robots is examined to provide the context for further discourse. The chapter ends with some concluding thoughts about future developments regarding the role of the human user in the shaping of emotionalised social robots.

9.2 Theoretical Concepts

Central to the theoretical debate regarding the use of mobile phones has been its role in enabling the presentation of the self of the user. The symbolic interactionist approach of Goffman (1959) and Mead (1967) explored the differences between private and public behaviours. The dramaturgical concept Goffman developed highlights how humans interact as if performing on a stage. Aspects of their self are kept 'back stage' and only shared with close confidents, if at all, and 'front stage' behaviours are not withheld but form their public persona. Ling and Pedersen (2005) were among the first to publish studies of mobile phone users in which this backstage/frontstage tension was explored in the context of mobile phone use. They highlight how it is being used to manage and balance the tensions between the 'I' and the 'me'; the persona that individuals choose to show and present to others is the one that manifests how they wish to be seen—the 'me'. But the mobile phone also allows its user to explore their self, the 'I', without having to communicate with others at all. In my prior studies (Vincent 2010) I found that mobile phone users share a special intimacy with their device and this extraordinary relationship is, in part, explained by the emotions involved as they explore their self and their close relationships. They do this by choosing the content that they keep on their phone, particular photos, messages, favourites, apps and so on. These are the unique indicators of their emotional memories and expressions of facets of their self. This personalised content is the sum of their identity and the information they draw upon to express their self to others.

This emotional interaction between humans and machines is the subject of Vincent's and Fortunati's (2009) work on electronic emotion. These are the emotions created, lived and relived as a result of interacting with machines; electronic emotions are not new or different from those experienced in all aspects of everyday life but they are prompted or stimulated by machines. Emotions are educed when humans and machines interact; as well as touching and using machines, electronic emotions could also be prompted by just thinking about them—you do not have to touch in order to sense or feel an emotion (Maldonado 2003). Knowing a memory, an image, a contact, a text message from a friend, is stored on the device is sufficient to stir emotions and the constant touch, presence or even thought, of the mobile phone could trigger these memories and emotions at any time. These special qualities of mobile phone use that combine human and machine capabilities are also explored by Sugiyama (2013) who posits that the melding of relationships and mobile phone use has turned the device into a quasi-social robot leading to a paradox of power and weakness caused by the 'heightened complexity in the relational dynamics' (2013 p. 82) that is involved in managing the equipoise between the different emotional, social and technological pressures the relationship entails. Note here that although the examples thus far regarding electronic emotions have been about close links between mobile phone and user, the principle that electronic emotions are created as a result of interaction with a machine also applies to all types social robots. However, emotion research regarding robots has tended more towards the creation of a robot that can feel emotions, interact with humans and react to situations in a totally human-like way (Breazeal 2003; Novikova et al. 2014). I move on now to discuss the background behind the development of mobile phones (any of which, when combined with their user, can be an emotionalised social robot) and the simultaneous development of social robots.

9.3 Background

Mobile phones are one of the most contemporary communications media and they follow a history of centuries of innovation and technological advance in human's understanding and development of computational machines. They also reflect changing communication practices that enable people to be in touch almost wherever they are in the world (Fortunati et al. 2012). Over the three decades since hand-held mobile phones began to appear in everyday use in Western Europe they have provided a highly individualised and intimate personal communications facility. An individually assigned phone number, and services and features that can be uniquely personalised have enabled the mobile phone to become a virtual icon of their user (Vincent 2003). The mobile phone is also different from other simultaneously available electronic computational information and communications devices (for example personal computers, laptops, etc.) due to its small size, portability and singularity.

The evolution of telecommunications and media services over the last two centuries has been associated with technological advances, societal changes and developments in day-to-day life (Vincent 2014). Each generation of new information and communications technology (ICT) brings with it new devices and the world of mobile communications is no different; the humble mobile phone, also called a portable phone, a cellphone or a handy depending upon where in the world you are, has moved on from the original 'brick phones' of the 1980s. In the mid-2010s it is now beginning to be known more generically as a 'smartphone',[1] although this is a particular category of mobile phone. There are thousands of models, makes and types of mobile phone from those that provide only basic voice

[1]In this chapter I use the generic term 'mobile phone'.

and text services (e.g. Doro PhoneEasy 508[2]) to others that have 'smart' connectivity and convergent capabilities that enable it to function as a fully wireless (WiFi) enabled personal computer (e.g. iPhone6[3]). There have been many steps along the way as the mobile phone transitioned from a simple analogue device that only made and received voice calls (c1975) to one with a digital interface on which data could be transmitted and thus text messages (c1991). A few years later camera phones were introduced and then feature phones which added access to the internet. Smartphones, originally launched in 2007, have higher data speeds, WiFi and convergent technology connectivity and, of course, there are many hybrids of these various mobile phone types with different combinations of capabilities.

Since 2013 mobile phones have further advanced in their form and functionality providing the option of mobile phones, that primarily use the 3G and 4G cellular technologies for connectivity, and tablet devices, that primarily used WiFi connectivity; both types of device can mostly also use both technologies. Voice communication over the internet is now common practice and so the original hand held mobile phones of the 1980s have evolved into a device with multiple interfaces and modes of connectivity. Thus many of the capabilities of mobile phones are now replicated on tablet devices and the convergence of devices and technologies continues apace.

This now means the mobile device in your pocket can have the capacity and functionality of a desktop computer, video and audio recording device, voice and text communicator with messaging services, radio and television, games console, music player, painting and art box, camera, newspaper, book, access point for multiple applications offering health monitoring, translation, education, photo editing, information access, document storage and more. This is the device that is in the hands of new mobile phone owners and users of all ages across the globe and which has enabled people to adapt and change their social practices; learn new digital skills associated with, for example, art and photography, or maintain and monitor their health to name but a few. It is important to note that the concept of the combination of user and their mobile phone creating an emotionalised social robot is one that applies to any type of mobile phone, and, to some extent, also to the tablets and tiny laptops that many people now carry as their primary communicator. The key is that it must be the primary portable ICT device on which the user is most reliant.

During this period of phenomenal mobile phone growth robots have also experienced an upsurge in technological advancement and usage, both as devices designed for niche purposes that at times supplant human actions, as well as those that support or complement them. Similar to media and communications technologies robotics and automatons have a long history as described by Fortunati (2013). Today robotics and autonomous systems are integral to future technology

[2]http://www.doro.co.uk/Products/Mobile-phones-and-accessories/Doro-PhoneEasy-508-UK/ (accessed 19 November 2014).

[3]http://www.apple.com/uk/iphone-6/?cid=wwa-uk-kwg-iphone-com (accessed 19 November 2014).

strategies globally. They are designed to support all industries, as well as being integral to future visions of an 'Internet of Things' (Ferber 2013) in which the machine to machine interaction exceeds human internet and wireless use and society becomes more dependent on ICTs and Robots to survive. Fortunati observes that robots have moved through developmental changes from early automata, mostly serving decorative and entertainment functions, to those having more critical contributions to society such as military robots prosecuting war or, in the domestic sphere, where they supplant traditional caring roles, especially those carried out by women. '*The robots have moved on to become "social" robots and as such they aim to substitute at least some parts of human caring such as company, affect, communication, and entertainment*' (Fortunati 2013, p. 121).

The purposing of a robot so that it can in some way manage affect is an interesting development, especially in the field of health and social care, and education. These are not robots that 'feel emotions' but rather ones that are designed to respond to or elicit an emotional response from their user as part of their functionality. To some extent humanoid, or animal robots such as AIBO[4] (a robot dog) and Paro[5] (a robot seal) are being made emotional as part of their being anthropomorphised by their human interactants—perhaps to make their use more comfortable and acceptable to humans. Some robots are programmed to respond to emotion signs and deal with the cognitive aspects of behaviours. Paro, the robot seal, acts a comforter for its user who is often a person isolated by loneliness or dementia. Paro is designed to respond to touch in some of the ways that might be experienced when petting a dog or cat[6] but it does not make too many demands of the user. The focus of the designers and makers of these robots is to interpret the human behaviours the robot will encounter and provide it with the technical capabilities to respond appropriately. Thus, for example KASPAR (Dautenhahn et al. 2009), a robot designed to interact with children with special needs, will react in a particular way such as if touched. KASPAR has been successful in the process of assisting children who have difficulty interacting with other people, however it responds to a particular action associated with an emotion rather than sensing and intervening before the child, for example, actually touches it. There are a multitude of examples of developments in robotic design such as Aldebaran's NAO and Honda's ASIMO[7] which are humanoid robots that mimic, and attempt to replicate, human interaction, movement and some emotion. Made to look like a traditional humanoid robot they have white body panels and mechanical hinged joints. In this description of

---

[4]http://www.sony-aibo.co.uk/ (accessed November 28 2014).

[5]http://www.parorobots.com/ (accessed November 28 2014).

[6]http://www.theguardian.com/society/2014/jul/08/paro-robot-seal-dementia-patients-nhs-japan (accessed November 28 2014).

[7]http://asimo.honda.com/ (accessed 3 December 2014).

NAO by Alderbaran there are some analogies with the mobile phone relationship already described and I will return to this further on.

> NAO is a 58-cm tall humanoid robot. He is small, cute and round. You can't help but love him! NAO is intended to be a friendly companion around the house. He moves, recognises you, hears you and even talks to you! Since his birth in 2006, he has been constantly evolving to please, amuse, understand and love you. In short, to one day become your friend. Aldebaran created NAO to be a true daily companion. He is the little creature who helps you be your best. His humanoid form and extreme interactivity make him really endearing and loveable. (Alderbaran[8])

Different from AIBO, Paro, NAO and ASIMO are the Geminoid robots developed by Ishiguro's Intelligent Robot Lab at Osaka University.[9] These robots are made with a skin-like body covering and machine interface to look, and be dressed, as lifelike humans. Plans for these human-like robots include roles as newsreaders or as receptionists in hotels where the check-in process is fully automated.[10]

Also from Ishiguro's team is an hybrid robot, the Hugvie, which translates into vibrations the voice of a caller to a mobile phone placed in a large humanoid shaped cushion that can be cuddled to feel the intonation of a loved one.

> The reproduced vibration that conveys heartbeats in alignment with that voice, all help to sense an active presence of the partner and intensify the affinity toward him/her. Although the transfer of information is merely through voice and vibration, it is an innovative communication medium that embodies the minimum required elements for transmitting a humanlike presence. (Hugvie Geminoid[11])

The difference between the Hugvie and the mobile phone as an emotionalised social robot is that Hugvie aims to deliver a heightened level of intimacy in relationships by simulating human presence in a physical form perhaps making it more like Sugiyama's (2013) description of a quasi-social robot. A mobile phone that is emotionalised by the electronic emotions it engenders in its user does not need the stimulation of hugging cushioned vibration. Hugvie is a simulacrum of the human presence, whereas an emotionalised social robot mobile phone creates and enables electronic emotions prompted by feelings and expression of the user's inner self ('I') and the shared self ('me'). Hugvie could be used to augment the presentation of self by adding additional sensorial experience to using a mobile phone, although it would need to be miniaturised if it was to be completely integrated and indeed future 5D and 5G technologies might well be the enablers to achieve this.

This short review of historical and contemporary developments in mobile phones and social robots highlights a growing tension between the technologies, their human users and the machine qualities; not least those which proffer solutions for activities that humans have not yet realised could be supported or replaced by robots.

---

[8]Who is NAO? http://www.aldebaran.com/en/humanoid-robot/nao-robot (accessed 3 December 2014).

[9]http://www.geminoid.jp/en/robots.html (accessed 3 December 2014).

[10]http://www.pcworld.idg.com.au/slideshow/564921/androids-will-greet-guests-japanese-smart-hotel/?image=3 (accessed 3 February 2015).

[11]http://www.geminoid.jp/projects/CREST/Hugvie.html (accessed 3 December 2014).

9.4 Discussion

It is clear from the background literature already explored that this relationship between human user and their *personalised social robot* (mobile phone) is different from cyborg technologies and robot machines, not least in that it was not created to be a social robot machine; instead it is the result of the symbiotic relationship of mobile phone and human emotions that makes this human machine interaction different. The outcome is a close relationship in which the device becomes a natural, innate, always available, taken for granted and assumed constant presence. This relationship relies on the adaptability of humans to accommodate the vagaries of technologies that do not always exactly meet needs but, nevertheless, can be accommodated. Kaerlein (2012; and this volume) has identified in his research an important point with regard to this always available connectivity and presence of the device: that of the immediacy that this *personalised social robot* mobile phone demands and can deliver. The function of a mobile phone as a mediator is now diminishing as there is, in effect, no need for mediation as the device and human are as one. Kaerlein examines the controversial Elfoid phone which looks almost foetus like rather than like a mobile phone, robot or other more conventional hand-held device. Designed by the Geminoid team the haptic qualities of the Elfoid skin-like covering is challenging; touch is an important part of the emotional experience but an artificial skin is possibly conveying a different kind of emotional response than real skin might. It is interesting to note also that although humanoid robots do not have these 'self' proclivities their human creators have focused on the public presentation of the robot 'self' in order to integrate them or at least make them acceptable to humans. Thus NAO has a 'cute' persona, a sweet voice and has learned to say 'ouch' when it falls. Its human interactants, or interlocutors, appear endeared to it much as they would a small child. What we do not yet know is what difference NAO and other similar robots make to the humans they interact with, and in particular the electronic emotions they engender.

Imbued within interactions enabled by the mobile phone are the emotions which are in human thoughts and actions. Furthermore, these emotional interactions are expressed on a mobile phone using more senses that on other devices as the constant touch and immediacy of contact explored above show how close the relationship between user and device has become. All this means that once we have a personalised mobile phone we can use it as a channel for emotions to provide emotional contact when needed or simply to feel the sense of a loved one's presence. We voraciously utilise it for the to and fro of emotional interaction both within the self and between humans. As humans become more familiar with their personalised mobile phones they use them more and more to complement their activities and, in turn, the machine becomes part of their emotional arsenal, their support. Witness the number of people using their mobile phone on a London Underground train where there is no WiFi or cellular signal; they are not communicating with others rather they are communicating with their self such as by reading downloaded books, emails, watching films, looking at photos, reading text messages and listening to music. In this way the mobile phone becomes a means of reflecting their feelings as they respond to content they have stored on their device; content placed there for a reason and which forms a part of their exploration of who they are, their identity.

Social robots are most likely to be considered to be robotic, machine like, repetitive, perhaps humanoid but in any event an inhuman thing created by humans to alleviate the strain, and tensions of particular tasks or perhaps to deal with life-threatening situations (harmful materials, bomb disposal, etc.). However, as humans see the benefits and the perhaps essential role robots may have in day-to-day living they start to have a different view of them. A robot cannot function without a human; even if this is merely turning it on and programming its original activities (and turning it off when not needed); there are an increasing number of instances where a human cannot complete certain actions without the intervention of a robotic interaction: a lift door opening in response to body movement; an electric door opening as we leave a building; a dishwasher that performs the task of cleaning and drying out crockery and cutlery or a washer drier machine that does the same for our clothes.

Where does the social robot end and the human being begin in this emotionalised relationship between humans and machines? When the machine is performing a distinct and pre-programmed function this is quite simple to explain, although there is a precursor to the physical robotic act which involves the human having a thought, thinking about how to execute it and then actually doing it. When we consider the mobile phone and its associated electronic emotions as an emotionalised social robot the distinction between human and machine becomes blurred. The thought might well be stimulated by the electronic emotions that are associated with the mobile phone—a special relationship, a message or image held on the device. Emotion is the glue that binds the mobile phone and its user together.

9.5 Concluding Thoughts

The emotionalised social robot discussed in this chapter is of the outcome of a symbiotic relationship between mobile phone and its human user. Whilst the interplay between a mobile phone and its owner may be similar to the experience of using a robotic device to fulfil particular tasks, there are exacting and extraordinary human and emotional differences that set it apart from this mechanistic view. To begin with the mobile phone was designed for the specific tasks of mediating voice and data between humans; now, it is designed to deliver a host of functionality and capabilities which the user can choose and shape to meet their own needs and thus make it their own. The result is a social robot that is uniquely created by iterative interaction between the same human and their own, personalised (and emotionalised) mobile phone. Each person's mobile phone becomes a one-off emotionalised social robot; there can be no duplicates as each device is personal and unique to its user. If access to another person's mobile phone is obtained the emotionalised social robot functionality cannot be transferred to a different user. The relationship between human and machine, the emotionalised social robot, has become completely independent of others.

What is clear from the parallel development of social robots and wireless communications is that we are heading towards an inevitable, and not so distant, convergence of these technologies and their uses. Rather than developing additional technologies it is quite likely that social robotics technologies may advance to deliver machines that do have emotions and are in other ways more intelligent than humans. Perhaps this is not necessary in all instances, however, as for example the mobile phone is already a miniaturised Hugvie because the vibrations we feel when we speak to our loved ones are felt in electronic emotions not in haptic interaction. The discussion draws me back again to my earlier assertion that the mobile phone and human are as one, the emotionalised social robot that combines mobile phone and its user is the synthesis of a highly technical machine that has capabilities often far beyond the wit of its user—a human user who has come to depend on the functionalities the mobile phone affords for the efficient running of their day-to-day life. The process of emotionalising the mobile phone and thus its role as a social robot has come about as a result of countless millions of human and machine interactions, each one involving electronic emotions and each adding to the personalisation and individualisation of the mobile phone, thereby intensifying its status as an emotionalised social robot.

References

Breazeal C (2003) Emotion and sociable humanoid robots. Int J Hum Comput Stud 29:119–155

Dautenhahn K (2004) Robots we like to live with?!—a developmental perspective on a personalized, life-long robot companion. In: Proceedings of the 2004 IEEE international workshop on robot and human interactive communication, Kurashiki, Okayama, 20–22 Sept 2004

Dautenhahn K, Nehaniv CL, Walters NL, Robins B, Kose-Bagci H, Mirza NA, Blow M (2009) KASPAR—a minimally expressive humanoid robot for human-robot interaction research. Appl Bionics Biomech Hum Robots 6(3–4):369–397

Ferber S (2013) How the Internet of things changes everything. Harvard business review innovation 7 May 2013. https://hbr.org/2013/05/how-the-internet-of-things-cha/. Accessed 28 Nov 2014

Ferrando F (2014) Is the post-human a post-woman? Cyborgs, robots, artificial intelligence and the futures of gender: a case study. Eur J Future Res 2:43. doi:10.1007/s40309-014-0043-8

Fortunati L (2013) Afterword: robot conceptualizations between continuity and innovation. Intervalla Platform Intellect Exch 1:116–129

Fortunati L, Pertierra R, Vincent J (2012) (eds) Migration, diaspora and information technologies in global societies. Routledge, New York

Goffman E (1959) The presentation of self in everyday life. Penguin Books Edition, Middlesex

Kaerlein T (2012) Presence in a pocket: phantasms of immediacy in Japanese mobile telepresence robotics. Communication +1 1(6). http://scholarworks.umass.edu/cpo/vol1/iss1/6. Accessed 30 Nov 2014

Ling R, Pederson P (eds) (2005) Mobile communications: renegotiation of the social sphere. Springer, London, pp 203–218

Mead GH (1967) Mind, self and society. Chicago University Press, University of Chicago
Maldonado T (2003) The body: artificialization and transparency. In: Fortunati L, Katz JE, Riccini R (eds) Mediating the human body: technology, communications and fashion. Lawrence Erlbaum Associates, Mahwah
Nitsch V, Popp M (2014) Emotions in robot psychology. Biol Cybern 108:621–629. doi:10.1007/s00422-014-0594-6
Novikova J, Watts L, Bryson J (2014) The role of emotions in inter-action selection. Interact Stud 15(2):216–223. doi:10.1075/is.15.2.10nov
Sugiyama S (2013) Melding with the self, melding with relational partners, and turning into a quasi-social robot: a Japanese case study of people's experiences of emotion and mobile devices. Intervalla Platform Intellect Exch 1:71–84
Syrdal DS, Koay KL, Walters ML, Dautenhahn K (2007) A personalized robot companion? The role of individual differences on spatial preferences in HRI scenarios. In: Proceedings of the 16th IEEE international conference on robot and human interactive communication WC3-2, Jeju, Korea, 26–29 Aug 2007
Vincent J (2003) Emotion and mobile phones. In: Nyiri K (ed) Communications in the 21st century mobile democracy essays on society, self and politics. Passagen, Vienna, pp 215–224
Vincent J (2006) Emotional attachment and mobile phones. Knowl Technol Policy 19(1):39–44
Vincent J (2009) Emotion, my mobile, my identity. In: Vincent J, Fortunati L (eds) Electronic emotion: the mediation of emotion via information communication technologies. Peter Lang, Oxford
Vincent J (2010) Living with mobile phones. In: Höflich JR, Kircher GF, Linke C, Schlote I (eds) Mobile media and the change of everyday life. Peter Lang, Berlin
Vincent J (2013) Is the mobile phone a personalized social robot? Intervalla Platform Intellect Exch 1:60–70
Vincent J (2014) What's so special about the mobile phone? exploring the mobile phone as a legacy of its ICT progenitors. In: Denison T, Sarrica T, Stillman L (eds) Theories and practice for community and social informatics. Monash University, Melbourne
Vincent J, Fortunati L (2009) Electronic emotion: the mediation of emotion via information and communication technologies. Peter Lang, Oxford

Chapter 10
The Technologicalization of Education in China: A Case Study of the Home–School Communication System

Chung-tai Cheng

This article regards the social implications of human–robot interaction in education in the specific sociocultural context of China. In particular the Home–School Communication System is taken as a case study. The first part of the article introduces how the standardized and automatized system helps student monitoring and school communication. The opinions of the different types of users involved (teachers, parents, and students) are then examined so as to look at the social impact of the system for care services and for communication services between teachers and parents. Through cross-cultural aspects concerning comparisons of education in Chinese and in Western societies, the article suggests that such kind of mediated communication may gradually change the role of teachers and parents in the Chinese context.

10.1 Introduction

With regard to the use of robot technology in education, people tend to think of "the robot as humanoid," a smart machine that attempts to imitate the human being and even replace human functions, such as being used to help children with autism or in the special needs classroom (Dreyfus 2009). Other studies concerning the power of robots focus on investigating how technological advancement may empower people so that the overall performance of education can be improved (Hiltz 1994). For example, Oppenheimer (1997) reports B.F. Skinner's suggestion that students are able to learn more efficiently with the help of teaching machines

C. Cheng (✉)
The Hong Kong Polytechnic University, Hong Kong, China
e-mail: cheng.chungtai@polyu.edu.hk

J. Vincent et al. (eds.), *Social Robots from a Human Perspective*,
DOI 10.1007/978-3-319-15672-9_10

and programmed instruction. In particular, with the development of the Internet, robot technology has shifted from the material capacity for the work of humans to many intangible aspects such as communication and emotion (Fortunati 2013). It seems the emergence of social robots in people's daily lives has been touted as a powerful force that will revitalize education through distance learning, particularly in remote areas. In other words, social robots are designed to make life easier in ways such as providing company, monitoring, and communication.

Concerning the quality of education, some scholars hold the opposite view and believe that good education needs face-to-face interaction between different stakeholders. Even though virtual classrooms are able to provide a sense of "community," real-life interactions cannot be replicated (Fu 2013). As N. Dye, the President of Oberlin College, asserts, "Learning is a deeply social process that requires time and face-to-face contact. That means professors interacting with students, while the Internet and the technology also play vital roles on campus (Paula Gordon Show 1999)." Both sides of the education/robot debate, however, presume that robot technology is only a handy tool, whose impacts are either positive or negative depending on how people make use of it. Since social robotics is being increasingly embedded into everyday life, this standoff may overlook the interplay between humans and robots in specific social contexts. This is because such human–robot interaction does not only refer to a platform for human interaction, but a medium with which humans interact (Zhao 2006). Thus, in face of the development of the automation process in the domestic sphere, we have to take a careful look not only at the role of social robots in daily life management but also how people perceive and understand technologicalization in everyday life (Haddon 2003).

In China, along with the rapid economic growth after the Open Door Policy, robotics has been appropriated from the industrial sector into different dimensions of public and domestic spaces. For example, many metro stations in cities have adopted ticket-vending machines; while most of the interprovincial train stations have kept ticket sellers. Given that the indigenous Chinese culture does not have a unique perception and understanding of the social contract between humans and robots, the adoption of robotic technology may generate some unexpected consequences in social life. In order to explore the arrival of robotics in Chinese everyday life, this article aims to explore the social implications of human–robot interaction in education in China, using a case study of the Home–School Communication System, also called Xiaoxuetong. First, the article will introduce how this communication system creates a standardized and automatized process mainly for student monitoring and school communication. Second, it will look at different stakeholders' views on the increasingly sociotechnological-based schooling environment. With the advancement of computer programming and mobile technology, the end of the article suggests that the emerging field of human–robot interaction in education may gradually change the roles of teachers and parents as well as the traditional Chinese understanding of "good education."

The study is one part of the large-scale project, "Social Consequences of Mobile Telephony in Mainland China," funded by the Department of Applied Social Sciences, The Hong Kong Polytechnic University, and conducted from

2008 to 2011. In addition to a national survey about mobile phone usage in six cities in China, the study conducted 20 interviews in each city, including Shanghai, Guangzhou, Dalian, Chengdu, Lanzhou, and Nanning, induced by snowball sampling. The discussion of the article is based on five in-depth interviews in Guangzhou, the capital of Guangdong province. All interview materials have been taped and transcribed.

10.2 School Management and the Home–School Communication System

In 2003, one of the largest mobile service companies, China Mobile, developed an SMS service plan, namely "Xiaoxuetong," as a new communicative platform for school communication between teachers and parents. But mobile phone usage in Guangdong province in 2003 was not very popular, with a penetration rate of around 40 % (Sina.com 2003). In 2006, the company provided a special offer and strategically cooperated with primary and secondary schools in Guangdong to further promote the service. The number of school partners increased from 1403 in 2006 to over 10,000 in 2010.[1] According to the company's half-yearly report in 2010, the service generated over 10 billion RMB in only 6 months. In 2011, this service had more than 45 million parent users, 2 million teacher users, and about 90,000 schools had adopted the system. In recent years, this big market opportunity has attracted several mobile service providers to further invest and invent different human–machine interfaces for education and e-learning services. The tendency toward the automation process in China's education has gradually attracted the concern of scholars and educational practitioners regarding the opportunities and challenges of the essence of quality education.

According to the company's mission and values statements, Xiaoxuetong aims to promote the all-round development of students in China and enhance their comprehensive ability and creativity through building a bridge for communication between teachers and parents and creating a digitalized campus. The company emphasizes that with the use of automation in education, teachers are able to save more time in handling administrative work and spend more time on the preparation of teaching materials; parents can know more about what their children do and learn in school, and students will receive a higher quality of education. The ideal is that Xiaoxuetong is not a simple communicative tool, but rather a sociotechnical educational landscape that integrates different institutions, departments, and groups of people in school through prescribed, automatized and standardized communications. This involves two different methods: First, Xiaoxuetong makes some practical parts of daily school management routines automatic, such as tracking the attendance of students and online meal booking and menu selection

[1]Please see http://tech.sina.com.cn/t/2011-10-25/08056225302.shtml.

for students; second, its central data management system also leads to the robotization of intangible parts of school communication and caring.

Generally, Xiaoxuetong's school partners need to set up a central data management system to store and manage students' personal information, including class performance and conduct, examination results, school fees, and related financial handling records etcetera. Students will also be given an IC card as their digital identity. When they arrive and leave school through the school main gate, the central system will automatically report their safety to the parents through sending an SMS, or a pop-up notification for smartphone users. With the help of the automatic notification system, parents will receive daily school news and be closer to updated school management arrangements, such as student daily homework notification, bad weather alerts and emergency notices.

In addition, Xiaoxuetong has provided several kinds of packages for service upgrading. School partners, teachers, and parents can buy additional services at a lower incremental cost from the company. For example, boarding school students are able to contact their parents by using the campus telephone service with the IC card; parents can also arrange online revision services for their children provided by partnering schoolteachers; teachers are able to freely contact students' parents either by using the group chat system or on an individual basis efficiently with the help of the automated voice messaging system. Looking at these services, the increasingly technological-based learning environment in China's primary and secondary schools reveals that robotic functions are no longer limited by the electronic entities, which have been embedded in everyday life "until they are indistinguishable" (Weiser 1991). In many cases, people may not even be aware of the presence of social robots, especially in terms of how they change the way people interact and communicate.

10.3 The Social Impacts of the Automatized Schooling Environment

Close Monitoring and Comprehensive Child-Care Services Since robotic functions have become more invisible, ubiquitous, and inseparable with other media, their social impact on our daily lives is almost imperceptible. When Xiaoxuetong was first brought to market, most people showed a positive attitude toward the "helpful product." Even today, while Xiaoxuetong's school partners, teachers, and parents apparently believe that the advancement of educational technology does help improve the quality of education, they generally regard the product as an instrument for communication only. In the study, most of the parent informants shared that Xiaoxuetong could promote more comprehensive child-care services:

> Parent A: Teachers (actually the central data management system) will send the kids' meal menu to us every day so that we're able to know what the kids eat at school… Once my youngest son got some red spots on his back, but he couldn't tell me what he had taken in

> the afternoon. Luckily, the service allowed me to check the meal list and help me sort out the source of his allergy.
>
> Parent B: My son always persuades me that he has grown up now. He said that he could go to school by himself as he was 10 years old. But I am quite worried that it may not be safe for him on the way back from school, and I think he is too young to have his own mobile phone. Fortunately, the Xiaoxuetong service will send me notification when he has arrived safely at school. I don't need to worry anymore.

The automatic safe arrival, emergency alert, and school notice systems really help school staff—particularly in primary schools—improve child monitoring. Most of the schools in China offer large-class teaching, in which each class has more than 40 students and at most only two class teachers. The student–teacher ratio and large-class size mean the teachers find it difficult to take care of all of the students. With the help of these systems, teachers really feel less worried about students' physical safety, and are able to complete various routine administrative tasks in a more efficient way. In addition, because of rapid urbanization after the economic reforms, working parents also love having such child-monitoring services available. For those rural laborers who have migrated to the coastal areas, the systems really help relieve their burden while they are dwelling in the cities far away from their "hometown" and family support.

Nevertheless, those families who have sufficient manpower to take care of their own children, tend to have negative comments about the Xiaoxuetong system and its charges. Some parents complain that they were forced to buy the services, saying they did not need it but had no choice. Some also complained that the automatic notification system was not reliable at all, and would even send some "useless" messages in the middle of the night (Sina.com 2011). As parent C shared:

> The system will send you some general information messages. You may receive nothing for a few days, but sometimes receive several messages at the same time—I don't know why, maybe it gets jammed. What's more, although the system helps teachers process and send out students' grades and the arrangement of homework, the teachers still need to input the data themselves. Just like in my son's class, if the teacher is responsible, you will get updated information. Otherwise, you might just receive useless messages.

In order to motivate teachers to fully utilize the services and render its services worthwhile, the company has implemented award/repayment schemes as incentives. For example, parents are allowed to check the total number of messages sent by different schools in the same area or the number of messages sent by individual teachers at the same school, online. These statistics are also listed and ranked on the Internet. Such an arrangement not only creates peer pressure amongst teachers, but it also becomes an informal criterion for parents as well as school management teams to grade and monitor teachers' performance. As class teachers may receive performance awards or bonuses from schools, and even from the company, some of them are eager to show they are working hard by sending messages for all kinds of school business.

Social Comparison Through Xiaoxuetong The unintended consequence of teachers' mass messaging is to foster unhealthy competition amongst parents and children. Parent C pointed out that apart from the automatic homework notification, some class teachers would demand parents' attention on students' performance by sending mass messages:

> 22/1/2009 [16:45]: Students have to finish two pages of the English copybook and one handwritten page and prepare for Chinese dictation for the coming Monday.
>
> 22/1/2009 [17:02]: Miss Chen found that Xiao Ding and Xiao Hui kept talking in the English lesson.
>
> 23/1/2009 [17:15]: Please remember to bring money for meal services tomorrow.
>
> 25/1/2009 [17:58]: Xiao Mui improved a lot in the Maths exam; while Xiao Tian and Xiao Ming received a fail grade.

Disclosure of student records to third parties is undoubtedly considered as a matter of personal privacy in Western societies. This is because children are regarded and treated as "individuals," whose personal feelings and rights need to be respected and protected. But the situation in China is more complicated. As mentioned above, in relational terms, the characteristics of Chinese social relationships are the opposite of the Western concept of the 'autonomous self'. Chinese people tend to be socially and psychologically dependent on others according to the differentiated and graded social relations. In Chinese social life, such interpersonal relationships based on particularistic criteria or ties, also known as "*guanxi*," are of paramount importance (King 1994). They not only serve for group and "individual" identification, but also generate different sets of moral guidelines for social interaction. Therefore, the practice of an in-class and examination performance notification system unexpectedly encourages intensive and even unrealistic "interpersonal comparisons" (in Chinese "*Panbi*") amongst parents and students.

In children's education, for example, Chinese parents have the rights and responsibilities to look after their children in all aspects of life. They will also try to provide what children need as best as they can. The most common way for them to show their love to their children is to work, rather than focusing on psychological and emotional support. On the other hand, children are expected to obey their parents, strive to meet their expectations and repay parents' kindness with filial piety in the future. As a Chinese classical text, *Book of Filial Piety* (in Chinese "*Xiaoji*"), suggests, "Our body, to every hair and bit of skin, is received by us from our parents, and we must not presume to injure or wound them. This is the beginning of filial piety." This means the ideal of this parent–child relationship in Chinese society is reliant and permanently based on reciprocal obligation. Under such family-oriented relationships, individual rights, and private emotions are usually insignificant. It is common for parents to search children's school bags, or freely enter their bedrooms without knocking in the name of "good intentions."

Unfortunately, in practice the family connection probably brings too much parental and family pressure on children. Since such *guanxi* is closely linked to "face" (in Chinese "*mianzi*") which corresponds to either a person's social image or the respect in which they are held, parents and other senior family members tend to set high expectations for children and neglect their feelings. As a Chinese saying goes, "Every family has its own problems and a sad tale to tell. Nobody wants his or her family's skeleton to fall out of the cupboard." If a Chinese child does something wrong or commits crimes, his/her parents may be accused of "bad parenting." As a result, for the teacher users, using Xiaoxuetong as a quasi-robot for mediated communication enables them to avoid uneasy embarrassment when they deal with difficult parents. As teacher A commented: "Sometimes it's hard to point out explicitly what sort of problems the students have. Actually, the reason why some students are slow to improve grades isn't due to the students' lack of interest, but to their parents' lack of time. Although parents may have their own difficulties, it's quite hard to remind them again and again in person." Teacher B shared the same view that some parents might feel sensitive and concerned when she sent written memos to them, while it seemed less formal to communicate with parents through Xiaoxuetong.

On the other hand, the practice of reporting students' examination results through Xiaoxuetong unexpectedly reinforces the overwhelming pragmatic concern in China's education, which has already been criticized for emphasizing marks and grades too heavily. Due to "face consciousness", Chinese parents and relatives are proud to share the honor and success of family members, such as getting good examination results, studying in top universities, making a lot of money, and other observable achievements. Although the automatized communication system helps teachers and parents closely monitor children at a lower cost, it creates a sociotechnological landscape that intensifies family and social pressure as "motivation drivers" to discipline and punish students for their poor performance. For example, student A commented, "I really suffer from a lot of pressure during the exam period. My parents work so hard for me. I really don't want my mum to feel disappointed." Student B also sighed that he was always warned by his mother that if he got poor examination results, she would cut his pocket money as a punishment for making her lose face.

These examples reveal that although Xiaoxuetong allows teachers and parents to communicate electronically, this new way of social interaction unintentionally reinforces the collective characteristics of Chinese social organization. As Chinese society's hierarchy is legitimate and an individual's conformity to group norms is significant in the family-based social provision, Chinese people are encouraged to identify their social position by comparing with others' material possessions (Chen and Prendergast 2007). Xiaoxuetong, therefore, is more than a communicative tool, as it not only involves the automatized process of school communication but establishes a new field for social comparison of goods in Chinese society. Previously, there was still room for both students and parents to hide "bad news"; while the networked school community makes it more likely that disclosure and close monitoring inevitably generate unnecessary and intangible pressure on students as well as excessive competition between parents. As a result, the learning environment is becoming more materialistic.

The Distortion of the Role of Teachers and the Value of Education in China Although Xiaoxuetong provides a cheap and efficient way of child monitoring for both large-class teachers and working parents, it unexpectedly reinforces the practice of social comparison amongst Chinese people alongside the development of the increasingly technological-based learning environment. At the level of social development, this phenomenon recalls a controversial discussion regarding the rationale of education in China, especially after the rapid economic growth over the past three decades. Since cultural practices are embedded and pervasive in everyday life, we can only feel and articulate their differences by contrasting the different styles with each other. In other words, our understanding of social robotics and its application in the educational system reflects different cultural styles; while the general cultural styles of education also condition how robotic automation in education is perceived and adopted. In education, many cultural anthropologists have studied and compared the differences of formal schooling and nonformal schooling between Chinese and Western societies, emphasizing that the Western one has a relatively clear distinction between the functions of school and of family.

In Western societies, children are generally expected to be independent and learn how to deal with different types of people in accordance with different social norms in different social institutions (Fei 1992). Schools themselves are thus perceived as one of the major organizations mainly for socialization (Hsu 1981). During the learning process, students are encouraged to develop their own distinctive personalities and learn how to interact well with each other. For example, American students are expected to express themselves and own ideas, and learn to take responsibility for their actions; while American teachers are like facilitators in the classroom, who encourage the students to do their own thinking. Gradually, students are able to learn their role and responsibilities in a variety of social circumstances and to recognize the difference between private and public life. In other words, the role of educational practitioners is to inspire and encourage the students to strive for their own greatness and potential. As Yeats (1996), the famous Irish poet said, "Education is not the filling of a pail but the lighting of a fire."

In Chinese society, however, the functions of family and school are not clearly defined and well structured. The schooling system is usually treated as an extension of the family regarding human development. Traditionally, it significantly focuses on the human subjective ethical spirit in enhancing family and then national cohesion. As the *Three Character Classic* (in Chinese "Sanzi Jing"), an ancient Chinese text of the embodiment of Confucianism for formal education at home, said, "When young, study; when grown up, perform. Influence the sovereign above, benefit the people below (Yellowbridge 2003)". The traditional ideal of Chinese education emphasizes group conformity so as to achieve a great united and harmonious society (in Chinese "*datong shehui*"). In education, therefore, Chinese people tend to have a unique and serious social perception toward the role of educational practitioners. As a Chinese saying said, "Once a teacher, you are a father figure for a whole lifetime." Teachers in China are usually trained to be "role models," being responsible for setting guidelines and instructions through lecture. Students not only have to learn to respect authority but to try to engage in a strong and long-term teacher–student relationship. Thus, Chinese education seldom emphasizes students' self-expression, and so the Chinese child tends to be more suppressed and attempts to tone down his or her own desire in order to transcend the greater things in life (Hsu 1983). Chinese students are expected to acquire knowledge through standardized, assigned and even "official" textbooks.

Through this cross-cultural comparison of education, we can have a deeper understanding of the co-construction of social robotics and culture in the Chinese context. Education in most cultural styles not only allows people to equip themselves by following some procedure or through imitation, but also teaches them how to acquire the style of their culture through socialization (Dreyfus 2009). The above sketch of Western and Chinese cultures of education suggests that the latter style seeks collective efficacy and social integration; while the former strives to develop individuality. Thus, although the promise of robotic communication technology enables school members to keep connected and build closer relationships with others, the style of Chinese culture governs how the technologicalized schooling environment is being perceived and shaped. Since Chinese society emphasizes collectivism and pragmatism, Xiaoxuetong is commonly regarded as a simple communicative device, particularly for teachers and parents. But if we look closely not only at students' comments but also at their understanding of the meaning of being "good" teachers, it is interesting to notice the normative implications of the automatized system of school communication. For example, student C commented,

> I hate Xiaoxuetong very much as my class teacher likes telling on me by sending "secret messages" to my mum. Once my friend and I were playing for fun, but she told my mum that I was fighting at school. I feel so aggrieved!

Contrarily, student D loved her teacher very much and said,

> All my classmates feel happy to have such a nice and considerate teacher because she never speaks badly of us but only sends good news to our parents.

These cases highlight how the Chinese evaluation of teachers is strengthened by the automatized communicative system. It seems Chinese students show less concern following the correct rules for promoting self-efficacy and one's own abilities. Rather, Chinese students' understanding of "good" teachers mainly focuses on the interpersonal characteristics of comprehensive caring and high compliance. Therefore, since Xiaoxuetong reinforces the practice of social comparison amongst parents as well as teachers, the role and responsibilities of "good" teachers may become exaggerated.

10.4 Concluding Remarks

With the advancement of broadband technology, various aspects of education are undergoing the processes of automatization and robotification in China. This study has argued that to understand the effectiveness and adoptability of robotification, it is important to look at users' understanding and interpretations toward the increasingly automatized and textually mediated social world. Especially in those developing countries, the interaction between technology and the indigenous culture may bring forth some unexpected social consequences. In China, the rapid growth of the economy after the Open Door policies contributes to a significant improvement in people's living conditions. The adoption of new information and communication technology that paves the way for social robotics has already spread to many domains of life. One of them is Xiaoxuetong, a matrix for school communication analyzed in this chapter, that helps teachers and parents better monitor students' performance. In terms of child caring, Xiaoxuetung has turned out to be a mostly affordable and practical tool to improve the efficiency of school management. However, its automatized communication features have unexpectedly encouraged the practice of social comparison between parents and teachers. Furthermore, such a sociotechnological network strengthens a culture of suppression and monitoring of students. This phenomenon reinforces the role of Chinese teachers in a more pragmatic style in school, and as a result may promote materialistic concerns in the Chinese educational system. As new automatized and robotized technologies are introduced in Chinese schools, these and other unexpected and potentially counterproductive effects should be taken into consideration.

References

Chen K, Prendergast G (2007) Materialism and social comparison among adolescents. Soc Behav Pers 35(2):213–228

Dreyfus HL (2009) On the internet. Routledge, London

Fei X (1992) From the soil: the foundations of Chinese society. University of California Press, Berkeley

Fortunati L (2013) Afterword: robot conceptualizations between continuity and innovation. Intervalla 1:1–14

Fu JS (2013) ICT in education: a critical literature review and its implications. Int J Educ Dev Inform Commun Technol 9(1):112–125

Haddon L (2003) Domestication and mobile technology. In: Katz JE (ed) Machines that become us: the social context of personal communication technology. Transaction Publishers, New Brunswick

Hiltz SR (1994) The virtual classroom: learning without limits via computer networks. Ablex Publishing Corporation, Norwood

Hsu FLK (1981) Americans and Chinese: passage to differences. University of Hawaii Press, Honolulu

Hsu FLK (1983) Rugged individualism reconsidered: essays in psychological anthropology. University of Tennessee Press, Knoxville

King AY (1994) Kuan-his and network building: a sociology interpretation. In: Tu WM (ed) The living tree: the changing meaning of being Chinese today. Stanford University Press, Stanford, pp 109–126

Oppenheimer T (1997) Comput delusion. Atlantic Mon 280:45–62

Paula Gordon Show (1999) The February 2000 nancy dye interview on the paula gordon show. http://www.oberlin.edu/news-info/00feb/paulagordonblurb.html. Accessed 12 Aug 2014

Sina.com (2003) The prevalent rate of mobile phone usage in China has increased more than that of the US and European countries. http://bj.sina.com.cn/art/20030326/32983.shtml. Accessed 12 Aug 2014

Sina.com (2011) Xiaoxuetong is accused of unreasonable charges. http://tech.sina.com.cn/t/2011-10-25/08056225302.shtml. Accessed 12 Aug 2014

Weiser M (1991) The computer for the 21st century. Sci Am 9:94–104
Yeats WB (1996) Selected poems and four plays of William Butler Yeats. Scribner Paperback Poetry, New York
Yellowbridge (2003) Three-character classic: a confucian roadmap for kids. http://www.yellowbridge.com/onlinelit/sanzijing5.php. Accessed 12 Aug 2014
Zhao S (2006) Humanoid social robots as a medium of communication. New Media Soc 8(3):401–419

Chapter 11
Fashion Tech and Robotics

Elda Danese

Many of the relatively recent experiments and realisations in the field of "fashion tech"—a combination of fashion design with engineering, science and interaction/user experience design—try to shape some sort of artificially intelligent systems around the human body, which in many cases move, breath and react to the environment around them. Although far from the social aim of health biorobotics, such researches have many elements in common with the latest advancements in the area of robotics related to the body, where the focus is often to enhance the wearability of the body-related devices, quite often through the employment of "smart textiles." The aim of this chapter is therefore to describe and analyse these parallel and partially intersecting developments, by presenting a review of the main projects recently realised by artists and designers who have collaborated with scientists and engineers towards the creation of "robotic dresses."

11.1 Introduction

Fashion has always had a close relationship with technology: The oldest and most obvious demonstration of this link lies in the constant innovation, since the eighteenth century, in the production of machinery and the implementation of processes for the manufacture of textiles. In recent years, new textiles have been produced and used in many different fields because of their lightness and strength, and also thanks to the fact that they could include various kinds of micro and nanotechnologies and incorporate conductive materials. In this field of research a new class of

E. Danese (✉)
IUAV University, Venice, Italy
e-mail: danese@iuav.it

J. Vincent et al. (eds.), *Social Robots from a Human Perspective*,
DOI 10.1007/978-3-319-15672-9_11

fabrics has emerged, the "smart textiles," which are engineered in order to react to changes in physical and environmental conditions. In many types of smart textiles, the fabric supports (or is made from) Phase Change Materials, electronic circuits, microcontrollers, sensors, etc. Inevitably, the experimentation in the sphere of "techno-textiles" is leading to the design of clothes that can be considered as peculiar kinds of machines.

Since worn fabrics are close to the human body, the field of "intelligent clothing" is advancing and expanding very rapidly, aiming to achieve new functional performances, chiefly in areas such as health, work, sport and the military. In this way the capabilities and the aspect of the body are changed, often improved, somehow like the performance of a machine would be. At the same time we have seen the development of many experimental projects, and some actual realisations, of robots trying to imitate the abilities and the appearance of human beings. Androids have a long tradition in the imagery and in the history of inventions: Recently, however, there can be detected a tendency to modify their artificial essence through the use of soft materials, therefore altering their metallic and geometric qualities to achieve more empathetic, naturalistic form. In these ways, as stressed by many scholars, the boundary between artificial and natural is more and more difficult to recognise (Danese 2003; Fortunati et al. 2003; Sugiyama and Vincent 2013). Moreover, this relationship has somewhat evolved, in the sense that technology is not only considered useful for functional tools, but it also has become fashionable: Hence the design of dresses that highlight artificiality in every direction, and that look spectacular because they appear closer to the inorganic and to the imagery of the machine. Finally, the interest of many researchers is increasingly focused on the emotive aspects of machines, studying the intimate relationship between their owner and them, designing their form in order to communicate, not just visually or phonetically, but through all the senses. From this point of view, as well as being close to the skin and communicating emotions, therefore, these new technological artefacts show an obvious association with fashion.

The main elements of the emerging field of "fashion tech"—a combination of fashion design with engineering, science and interaction/user experience design—can be exemplified in the projects carried out at the *V2 Institute for the Unstable Media* in Rotterdam, which in February 2014 hosted a lecture on "Robotic Fashion and Intimated Interfaces." A second exemplar can be found in the 2012 "Technosensual" exhibition at the Wiener Museumquartier, which looked at the world of technologically enhanced garments and raised questions on how such wearable technologies may impact the social, emotional and cultural layers of society.

With the aim of exploring these subjects, the chapter reviews some interesting research programmes, designs and actual realisations that in different ways incorporate elements related to the fields of fashion tech and robotics. The exploration of these themes has been structured into five sections: (a) Wearable robots, dealing with various soft exoskeletons projects, mostly aiming at supporting everyday activities of people with disabilities; (b) hiding/showing, where the main subject is the design of clothes or other "soft" equipment capable of generating some form of camouflage; (c) protection/aggression, covering the design of garments that enhance the degree of personal security and protection against various possible forms of aggression or attack; (d) communicate/express, discussing the development of garments capable of altering some of their characteristics in response to other people's presence or to the wearer's changing emotions; (e) shifting silhouettes, examining how some artists and fashion designers have interpreted the theme of "kinetic garments" or "transforming dresses."

11.2 Wearable Robots

The *Robot Companions for Citizens Manifesto*[1] reports some interesting new conceptualizations in the field of robotic technologies. One among them is "the 'Robot suit,' that is a wearable robot that provides support to people when moving and doing everyday life activities" (Fortunati 2013, p. 125). In recent years this area has seen various interesting research achievements and also a few actual realisations. The Japanese company Cyberdyne, for example, is producing a wearable robotic suit called HAL (Hybrid Assistive Limb), presented as the world's first cyborg-type robot—leading to a fusion of man, machine and information—by which a wearer's bodily functions can be highly improved, supported and enhanced. A similar device, manufactured by the Israeli company Argo Medical Technologies, is "ReWalk," a wearable robotic exoskeleton that provides powered hip and knee motion to enable individuals with Spinal Cord Injury (SCI) to stand upright and walk.

A number of other experimental studies and projects are being carried out around the world by research institutions and companies—frequently in partnership—in the field of wearable exoskeletons, often relying on different designs for their prototypes. These include: The Berkeley-based Ekso Bionics; the BioRobotics Institute of the "Scuola Superiore Sant'Anna di Pisa;" the American firm Raytheon Sarcos; the Biomechatronis Group at the MIT or the "Exo-Legs" project (coordinated by the University of Gävle in Sweden).

Of particular interest for the topic discussed in this article is the soft robotic exoskeleton suit recently developed at the Wyss Institute for Biologically Inspired Engineering of Harvard University. The lightweight "Soft Exosuit" is designed to overcome the challenges of traditional heavier exoskeleton systems, such as power-hungry battery packs and rigid components that can interfere with natural joint movement. It is made of soft, functional textiles woven together into a piece of

[1]The Robot Companions for Citizens (RCC), or RoboCom, is a large scale European interdisciplinary research and development initiative, whose aim is the creation of a new generation of robots that will co-exist and work together with humans. The whole programme is coordinated by Paolo Dario (director of the BioRobotics Institute of the "Scuola Superiore Sant'Anna di Pisa") and involves 96 principal investigators from 73 Institutions.

smart clothing that is pulled on like a pair of pants and is intended to be worn under a regular gear. Through a biologically inspired design, the suit mimics the action of the leg muscles and tendons when a person walks, and provides small but carefully timed assistance at the joints of the leg without restricting the wearer's movement.[2]

Most of the previous examples are related with problems affecting disabled people or with specific operational tasks that need an enhancement of the body's power: The possibility to move, to improve one's strength or to control some physical functions. While this is a crucial field of research, there are other aims that have emerged as specific applications of technology to garments. In particular, all the examples described in this paper demonstrate a connection with the communication aspects of clothing or—and this is another important subject—with what could be interpreted as its opposite, that is the wish to hide, the desire of not being controlled.

11.3 Hiding/Showing

What can be seen as the obsessive and increasing presence of systems of video surveillance in public spaces could generate, in some people, a willingness to escape from this form of exercise of power, to become actually invisible. Camouflage clothing has typically relied on the use of colour and patterns to conceal the person wearing it. By exploring in depth the field of "optical camouflage," a Tokyo University team led by Prof. Susumu Tachi[3] has developed the "invisibility cloak," a coat that seems to be transparent and appears to make its wearer invisible. Designed mostly for medical and military applications, the "invisibility cloak" is in fact a screen in which the images shot by a camera placed behind the person wearing it are projected on the front of the coat, thus giving the impression of seeing through the body. The use of a technique called Retro-reflective Projection Technology (RPT), which employs materials covered with very small beads that reflect the light only in the same direction as it is coming (the same type of material applied to road signs), makes the projected image bright and clearly visible even in daylight.[4]

Another illustration of the development of clothes that hide the identity of the wearer from invisible and ubiquitous eyes is represented by the products of the English design company Vexed Generation, that started its first collection in 1995. Its aim was to design a range of street clothing that met both the practical needs and political concerns of the "urban generation," expressing concerns over the

[2]See image at: http://www.slashgear.com/soft-exosuit-gets-darpa-favor-will-give-soldiers-super-human-legs-12346076/. Accessed 15 Jan 2015.

[3]Founding director and fellow of the Robotics Society of Japan, professor Tachi's research encompasses robotics, augmented reality, virtual reality, and haptics.

[4]See image at: http://www.techeblog.com/index.php/tech-gadget/scientists-develop-light-bending-invisibility-cloak-that-works-across-several-frequencies. Accessed 15 Jan 2015.

prevalent use of video surveillance: Their "Ninja Hood" and "Ninja High-Neck" tops, for example, helped conceal the identity of the people who would wear them. In their "See And Be Seen" Collection, instead, the approach was to enforce the garments' visibility in relation to the wearer's anonymity by inserting high-reflective panels into their overcoat and full body jumpsuits: The reflective panels shone as bright and fluorescent lights and thus masked the wearer even more fully in the contrasting folds of dark fabric, while extra padding reinforced the elbows and knees in the event of an accident or an altercation.

11.4 Protection/Aggression

Clothes typically protect us, define our identity and communicate to the others. Clothes, however, can also defend us from attacks, not just in wars or combat situations, but also from the unexpected dangers that people fear in modern city, the aggression of chemical substances in the air and the violence of unknown people. To many people, perceptions of rising crime, inner-city violence and the threat of terrorism make protection a key concern. For many years scientists and researchers have been developing textiles that create protective clothing for a range of applications, and many of these innovations have the potential to form part of the wardrobe of the future. Dutch designer Tim Smit, for example, created the "Urban Security Suit" to insulate the wearer against pollution, airborne toxins and poisonous gases: The garment is made with a Neoprene shell for insulation and shock protection, and is lined with body-moulded slash-proof Kevlar that protects the wearer from attack.

A more active aspect of the above-mentioned concerns, on the other hand, is exemplified by the aggressive function assigned to some clothes that are specifically designed to harm the people that try to touch us or hurt us. Different types of technological systems can be integrated into garments in order to obtain specific effects in such a direction: Soft circuits, conductive fibres and pressure-sensitive panels make it possible, for instance, to create clothing that can react and respond to impact and force. MIT researcher Adam Whitton has partnered with fashion designer Yolita Nugent to focus on personal security issues affecting especially women, devising items of clothing that sense contact and register the amount of force used. That led to the development of the "No Contact Jacket" which—when turned on through an internal switch—creates on the external surface an electric field capable of delivering a high-intensity non-lethal charge that causes any attacker to immediately withdraw contact, while simultaneously protecting the wearer from any electrical charge and allowing him or her time to escape.

In a similar but definitely more harmless way, the "Spike Jacket," designed by Nancy Tilbury, aims at amplifying the wearer's perception of his or her personal space: The jacket's integrated technology system senses when other people come too close, immediately prompting a system of textile cabling and silicone light diffusers to flash repeatedly as a form of warning signal.

11.5 Communicate/Express

Another quite active line of research focuses on the possibility for the garment to cause surprise or to express emotions by changing their passive status of objects. Therefore, their fabric may light up, move, swell up, become a screen or release perfumes. The aim seems to be the improving capability of our clothes to change their status in order to communicate to the senses.

Barbara Layne is the Director of "Studio subTela" at the Hexagram Institute of Concordia University (Canada), which focuses on the development of intelligent cloth structures for the creation of artistic, performing and functional textiles. Natural materials are woven in alongside microcomputers and sensors to create surfaces that are receptive and responsive to external stimuli. In 2007 the Studio realised "Jacket Antics," two garments where traditional black linen yarns are woven alongside light emitting diodes, microcontrollers and sensors. Their distinctive characteristic is that they rely on the act of holding hands by two wearers in order to present a range of dynamically encoded texts and designs scrolling through the LED array on each of the backs. If the wearers do hold hands, the LED arrays presents a synchronous message that scrolls from one person to the other, but when the wearers let go of their hands, the message changes to individual and different themes. In these animated cloth displays, the capacity for interactivity extends the narrative qualities of fabric and provides new possibilities for dynamic social interactions.

As wearable systems identify bodily sensations and trigger responses, some emotions could be processed as a type of computation that translates into action. Philips Design was one of the first research organisations to combine wearable technology and sensory intelligence. Among the various prototypes of dynamic garments developed in its laboratories one can cite the "SKIN Bubelle," described as an "exploration into emotional sensing." The garment is formed by a series of bubble-like shapes that glow individually at an intensity related to the wearer's movements and changes in skin temperature. To that purpose, a series of biometric sensors collect data such as hearth rate, respiration and galvanic skin response, and this information is then visualised by altering the intensity, shape and colours generated by 18 miniature projectors located between the layers of the garment. The effect creates a visual representation of the wearer's emotional state and physical responses.[5]

In an analogous project, Korean designer and researcher Eunjeong Jeon has brought natural fibres and technology together to create "Trans-For-M-otion," a conceptual prototype of a 'kinetic garment' that morphs into different silhouettes. As the designer states, the prototype "is based on investigating the bodily aspects of people's interaction with clothing, … to understand four basic emotional movements: 'towards out,' 'toward self,' 'away from self (against other)' and 'away

[5]See video at: http://vimeo.com/32964255. Accessed 15 Jan 2015.

from other'." (Jeon 2009). The outfit is crafted from offcuts of felted wool embedded with wearable technology, and it is designed with regular three-dimensional geometric polygon-shaped units, with air trapped on one side and flower-shaped pockets on the other. Sensors monitor muscle tension, breathing, heart rate and temperature to detect when the wearer experiences discomfort, and the garment reacts automatically by closing around the wearer to foster a greater sense of security. As a result, the shape of the silhouette relates directly to wearers' sense of emotional and physical well-being.

Jenny Tillotson—reader in Sensory Fashion at Central St. Martins College of Art and Design in London—is a pioneer of wearable, scent-output systems and aroma technology, whose interests centre on finding novel ways of creating mood-enhancing responsive fashion and textiles, using the olfactory sense as a communication tool. One of her experimental works, the "Smart Second Skin Dress," also known as "Scentorgan Dress," is a prototype garment that delivers aromas to different parts of the body. Like the human body, the garment has its own circulation systems, being cabled with "veins"—in this instance, medical tubing fastened at the end with surgical clamps—which hold coloured liquids of different aromas. A small pump mimics the function of the heart and blood vessels to emit scents that create and enhance the wearer's olfactory experience. In this and other analogue projects, "the aim is to go one step beyond passive sensory systems … via the integration of wearable technologies in smart textiles that not only offer function to fashion, but are designed for psychological end benefit to reduce stress" (Oliver et al. 2009, p. 10).

11.6 Shifting Silhouettes

Sweden-based fashion designer Julia Krantz's inspiration seems to emerge from biomimicry, robotics, genetical modification and combat uniforms. She created her "Shell Collection" with garments comprising translucent fabrics draped over metal frames, shaping the fashioned body into otherworldly silhouettes, even giving it an insectoid appearance. In another garment by the same designer, dubbed "Whiteness," the body is covered in rigid panels that suggest robotic body parts more than traditional garments. Tube-like structures tracing the shoulders and torso suggest technological circuitry, while the overlapping panels appear ready to morph into new shapes.

Australian born Lucy McRae is an artist and designer whose works reconfigure the shape of the human body. Taking the human form as her starting point, McRae designs wearable artefacts that dramatically transform the body's natural silhouette, and create coatings that resurface the skin. By using textiles and other fibre-based forms, or low-tech substances such as foam, feathers, paper and wood, McRae crafts wearable structures for the body to inhabit. A typical example of such kind of structures, transforming the body into sculptural shapes, is "Transnatural," a honeycomb-like textile mesh, made from the same thermoplastic employed for masks restraining the patients' body parts during MRI scanning.

The latest realisation of Ying Gao, fashion designer and professor at Université du Québec à Montréal, has been a series of kinetic garments, whose aesthetic attributes change and adjust in reaction to sound: The "Incertitudes" series, consisting of dresses in white and silver coloured fabric covered with dressmaker pins. The garment is realised through the combined use of quite different mediums and technologies: PVDF, a thermoplastic fluoropolymer having a strong degree of piezoelectricity, which makes the material compress when exposed to an electric field; a series of microelectronic devices, fully integrated into the fabric; thousands of dressmaker pins, outwardly protruding from the textile's surface. Through the interaction of these components, the whole dress and its metallic accessories move and respond to the level of noise and the voice of people in their surroundings. Their fluent motion generates a wave-like flux, contracting and expanding the entire wearable object.[6] One can say that in this case the clothing's original function is transposed into a unique aesthetic application, embodying art, fashion and technology. Seeing the reaction of the dress to the sound induces emotion giving the illusion of an effective communication.

Diana Eng's 2010 "Fairytale Fashion" is a collection of clothing that uses technology, math, and science to create functioning designs that transform shape and change colour. Her research into integrating *deployable structures* (i.e., structures that can change shape so as to significantly change their size) within fashion is exemplified in the "Inflatable Dress," made from cream silk chiffon draped over inflatable plastic forms and white silk flowers: When the inflatable are turned on, the dress shows a complete and impressive transformation.

Such a project clearly brings to mind one of the proposals put forward by the famous fashion designer Hussein Chalayan, who has always been interested in the movement and transformation of garments: In his 2003 collection "Kinship Journeys," in particular, he presented a series of dresses where inflating airbags emerged from the skirts' hems. As an immigrant from Cyprus to London, Chalayan is quite sensitive to the subjects of transit, exile and nomadism, and many of his collections make reference to voyage and travel, particularly air travel. In his "Echoform" collection (1999), for example, he presented some leather garments inspired by car interiors to represent speed, adding padded headrests to the dresses using the same technology as in car manufacturing. His "Aeroplane" series (2000), on the other hand, was designed using the same composite technology used by aircraft engineers. Glass fibre and resin were moulded into two smooth, glossy, pink-coloured front and back panels that fastened together by metal clips.[7] Referring to a quite similar outfit presented by Chalayan in the same year, where

[6]See image at: http://inspirationist.net/incertitudes-sound-activated-clothing-by-ying-gao/. Accessed 15 Jan 2015.

[7]See image at: http://www.metmuseum.org/toah/works-of-art/2006.251a-c. Accessed 15 Jan 2015.

the rigid side and rear flaps could be lifted, by means of a remote control, to reveal a froth of pink tulle underneath, Quinn (2002) wrote:

> The structural architecture of the Remote Control Dress echoes the attributes of a fashioned body rather than an organic body. The structure of the dress forms an exoskeleton around the body incorporating elements of body consciousness … As the dress interacts with its immediate environment, or performs manoeuvres originating from a command centre … makes it clear that the fashioned mechanisation of the body and the interaction of both into a larger technological system produces a whole new range of practices, possibilities and aesthetics that transgresses the body/machine boundary. (Quinn 2002, pp. 367–368).

In a correlated proposal, presented in the same year, some dresses incorporated a seat, a headrest and two armrests, all connected to a metallic spine on the back, creating a sort of prosthesis to the body. Finally, in his 2007 "One Hundred and Eleven" collection many suggestions were made in the area of "transforming dresses," where complex computerised systems of tubes, wires and motors, attached to the outer layer of the garment, were utilised in order to completely alter its shape.

11.7 Final Remarks

In summing up the preceding sections, the first aspect to be pointed out is that very few of the these "robotic dresses" have been conceived for being normally worn and even less to be put into production: Some of them can be considered as products being still at an experimental/preliminary stage, while others can be seen as some sort of artistic or stylistic propositions. As Lamontagne writes, referring to the projects presented at the "Technosensual" exhibition, "very few of the featured garments could function aesthetically or mechanically off the catwalk, or outside a gallery setting." Nevertheless, they all "raise deeper questions about our relationship to technology as it ubiquitously and seamlessly bleeds into objects of quotidian interaction, exchange and expression." (Lamontagne 2012, p. 15).

In the projects of many of the mentioned designers, in fact, the technology acquires many visual, sensory and emotional functions. The artificial light that radiate from many garments is a technological quality that transforms the body in a luminous immaterial essence. In some cases, the complexity of the pattern and the presence of pipes and rigid scaffolding evoke inorganic forms or the exoskeleton of an animal. It is important to stress the fact that, from this point of view, there seems to be a sort of exchange between the traditional image of fashion, on the one hand, and the most advanced robotic technology: The latter seems to point mostly towards the organic and human aspects, while the described kind of fashion tends to relinquish the supple and flat surfaces in favour of three-dimensional, sometimes sharp-edged shapes, pointing generally to some sort of non-human or post-human representation.

Finally, most of the reviewed garments, as we have shown, make use of new technologies, employ "intelligent textiles" and are equipped with advanced components in such a way as to become sophisticated systems (or machines) that react to/interact with the wearer but also with other people's presence, observation, talk, touch, etc. Therefore, by rephrasing and modifying a frequently cited characterization of social robots expressed by Shanyang Zhao,[8] we could say that most of these new "robotic dresses" are *both* a medium with which humans interact *and* a medium through which humans interact.

References

Danese E (2003) Soft machine. In: Katz JE (ed) Machines that become us: the social context of personal communication technology. Transaction Publishers, New Brunswick, pp 267–276

Fortunati L, Katz JE, Riccini R (eds) (2003) Mediating the human body: technology, communication, and fashion. Laurence Erlbaum, Mahwah

Fortunati L (2013) Afterword: robot conceptualizations between continuity and innovation. In: Sugiyama S, Vincent J (eds) Social robots and emotion: transcending the boundary between humans and ICTs. Intervalla 1, pp 116–129

Jeon E (2009) Object playing with movement: a source of comfort and enjoyment. Internal Paper, Department of Design, Curtin University of Technology, Perth, Australia

Lamontagne V (2012) Exhibition review. In: Wipprecht A, Friesinger G (eds) Technosensual: where fashion meets technology. Catalogue of the exhibition at the MuseumsQuartier, Edition Mono/Monochrom, Wien, pp 15–17

Oliver R, Tillotson J, Toomey A (2009) Bioengineered textiles and nonwovens—the convergence of bio-miniaturisation and electroactive conductive polymers for assistive healthcare, portable power and design-led wearable technology. J Fiber Bioeng Infomatics 2(1):1–13

Quinn B (2002) A note: Hussein Chalayan. Fashion technology. Fashion Theory 6(4):359–368

Sugiyama S, Vincent J (eds) (2013) Social robots and emotion: transcending the boundary between humans and ICTs, Intervalla 1

Zhao S (2006) Humanoid social robots as a medium of communication. New Media Soc 8(3):401–419

[8]"Humanoid social robots … are not a medium through which humans interact, but rather a medium with which humans interact" (Zhao 2006, p. 402).

Chapter 12
Conclusions

Jane Vincent, Sakari Taipale, Bartolomeo Sapio, Giuseppe Lugano and Leopoldina Fortunati

Although the topic of social robots has enjoyed great success in scholarly literature since 2000, this book contributes new knowledge to the debate that has been generated from diverse sources. It includes authors who come from many different countries delivering robust, quantitative studies on the attitudes and perceptions towards robots in Europe and North America together with a series of more qualitative research projects and experiments from Europe and Asia. Furthermore, it includes a good mix of disciplines: from communication to social policy, from sociology to industrial design, from philosophy to social psychology. Collaborating together all these disciplines have delivered integrated knowledge and a variety of studies regarding social robots.

In particular this volume has introduced and articulated a sociopolitical analysis of the penetration of social robots in the reproduction sphere, and in so doing

J. Vincent (✉)
University of Surrey, Surrey, UK
e-mail: j.vincent@surrey.ac.uk

S. Taipale
University of Jyväskylä, Jyväskylä, Finland
e-mail: sakari.taipale@jyu.fi

B. Sapio
Fondazione Ugo Bordoni, Rome, Italy
e-mail: bart@fub.it

G. Lugano
COST, Brussels, Belgium
e-mail: Giuseppe.Lugano@cost.eu

L. Fortunati
University of Udine, Udine, Italy
e-mail: fortunati.deluca@tin.it

J. Vincent et al. (eds.), *Social Robots from a Human Perspective*,
DOI 10.1007/978-3-319-15672-9_12

offers the beginning of an intense dialogue between policy makers and social scientists engaged in this field of research. It adds many interesting insights about the processes of robotification experienced in our contemporary society.

At the heart of the debate in this volume is the tension between human and machine that manifests in the question: in contemporary society how acceptable are human behaviours in robots and robot-like behaviours in humans? This interplay between machine and human, the to and fro of the robotic turn is exemplified in the analysis we have read herein.

The discourses are framed by the recent large survey from Europe (Eurobarometer 2012) which showed that people's ideas about what the robot is corresponds more with the picture of an autonomous industrial robot than with a human-like robot that could help people in their daily chores at home (see Taipale et al. in this volume). This reinforces the understanding that what people primarily consider a robot is something already used in industry, not something (more or less futuristic) that is being developed on designers' desks and in robot labs. However, at the same time we also learned that in robot labs many social robots are indeed being developed that push the boundaries of these conventional views. It is not uncommon for roboticists to draw on human interaction with animals, and dogs in particular, to inform their research regarding the relationship and (emotional) bonds between human and machine (Dautenhahn 2004), as well as research that informs the robots designed for health care applications (e.g. Paro) and emotion management by children with special needs, such as autism (KASPAR). Examples of robots discussed in this volume include animal, as well as humanoid robots but what more can we learn from observing and understanding humans in their everyday lives? The human interaction with social robots that we have explored is not simply that which deals with the reproduction sphere; we learn from Danese, for example, that there are a multitude of projects delivering wearable technologies that can transform the person clothed in them into a robot lookalike, give them cyborg-like properties to protect them from harm, or even act as aggressor. Furthermore, Vincent argues that the emotional needs of everyday life mediated via the combination of personalised mobile phones and their human user has created millions of individually tailored, unique, mobile emotionalised social robots—robots created at the hand of their user and not by technicians and engineers for their user.

These contemporary examples of social robots exemplify the historically and culturally shaped images of and attitudes towards robots on the one hand, and actual human behaviour with robots on the other. Höflich and El Bayed's chapter in this book shows that the people implicated in their research commonly picture robots as a cartoon and box-like character, reflecting the strong influence of media and culture on the understanding of what robots look like, but these robots are foreigners who are expected to come and take our jobs. These studies about actual human–robot interactions show that people are really curious about robots. Höflich and El Bayed further show that robots are actively engaged in social interactions with humans, they are welcomed as third members in human–human interactions, and in these interactions robots become quickly anthropomorphised by human users/operators.

With this book we have shown that our stereotypical images of the potential users of robots are largely informed, both by our experiences of the early adopters of other technologies and by distorted images repeatedly conveyed through media. For example, Taipale et al. revealed in their chapter that social groups, which are easily labelled as technologically non-savvy or the "laggard" adopters of new technology, such as pensioners, are, contrary to normative expectations, actually the most interested in having robots in the life domains that in general encounter most opposition among the whole population. Furthermore, Law's chapter reveals several sociopolitical factors that influence the robotification of contemporary Hong Kong. As long as mechanical aids for the disabled are designed only to serve physical and functional recovery and a disabled person's social needs are dismissed, these devices remain just mechanical robots and fail to develop into social robots. Law demonstrates how political decisions and economic rationalities affect this process by compelling many to continue to use old, less sociable, less automatised and less roboticised technical aids.

Against this backdrop the alternative approach to the study of social robots that was presented in the introduction of this book speaks for itself. By placing the human in the centre of the analysis we have revealed that humans as social, cultural and political actors have strong stereotypical preconceptions of robots and of their users. However, and much more interestingly, in the end these preconceptions may not hinder the adoption and penetration of robots as much as we might expect. Lessons from other domestic technologies, especially from ICTs like the mobile phone, show that as a rule people are inquisitive about new devices, ready to test and try them, and willing to adopt and use them, especially when they find uses that serve their personal needs.

In the introduction we also suggested that by beginning from the "softer" technologies and applications that are already widely adopted and used by ordinary people and that in fact are the only way to make a plastic and metallic body of a robot *social* (interactive, communicative, emotional, etc.), we could better understand the ways robots might enter the private and domestic spaces. Katz et al.'s chapter provides solid support for this approach. The authors show how earlier experiences in using online communities and avatars pave the way for the adoption of robots. The robotification of the society does not take place overnight. Apart from the technological innovation and marketization of robots, it requires that people have been accustomed to the automatised functions and processes of everyday life before the marketization and adoption of social robots can take place in earnest. In this respect, social media technologies and online environments work as an unparalleled spring board for the adoption of social robots as the same technologies are being incorporated in robots to make them sociable and easy to use. Illustrative of this is Cheng's study that shows the automatised information management/processing between parents and teachers in Chinese schools. Implemented as a mobile application it manifests as a step towards the robotification of education in China. The messages of Cheng's study are that once people can be encouraged to be accustomed to and recognise the value of automatised processing of information in the different sectors of society, the same technology should be easier to combine with a robotic body. If the robot is developed in this order, its social functions should become more effortlessly legitimised as the mechanical robot body will perform the social/cultural/interactive task already regarded as meaningful and functional by users. Overall, these observations suggest that once again young people would have an edge over the older citizens in the taking up of a new technology. At the same time, this highlights the need to engage adults and older adults better in the world of social media and the Internet. As a large share of social robots is especially targeted to help ageing people, familiarity with social media will certainly pave their way to a more and more automatised and roboticised future as active citizens.

We have also learned more about how people react to what the robot look like, and in particular the question of human-likeness which is embedded in robot studies. This question was addressed from many perspectives in the book, keeping the human at the centre of analyses. Most of the studies of the book speak for some sort of simplification of robots. Kaerlein's study about the reduction of the complexity of robots goes to the heart of the simplification debate. As is made clear in this volume, instead of continuously increasing the complexity of robots with the ultimate aim of making them human-like, it might be a viable strategy to consider the specific needs of a user first and aim to develop robots with a simple design and a limited number of technical functions. Indeed, the ability of humans to recognise the positive values of machines in support of their day-to-day life is well articulated in Vincent's chapter in which the mobile phone is appropriated and made into an emotionalised social robot as a result of the constant emotional interactions between the user and (mobile phone) machine.

It is worth noting that hitherto all widely spread and successful home electronics and domestic devices have relatively simple designs and are usually simple to use. On this basis when users' technical needs for a robot are known maximising the amount of social and communicative diversity the robot supports, as opposed to its range of functions and technical properties, might appear to be a useful strategy for the commercialisation and marketisation of robots. We learned from Höflich and El Bayed's study that presents people's attitudes and perception of robots that in Germany people would be more likely to accept robots that are least human-like. Although this issue is inconsistent with the well-known, albeit disputed, uncanny valley thesis, it makes sense in their study context. The least human-like robot examined was a robotic vacuum cleaner, which has no resemblance at all with a human being, but which is easy to accept by ordinary people because of its similarity to other domestic appliances. When robots have little or no similarity to home electronics or other domestic appliances, small likeness to human beings may hinder people's willingness to adopt robots more than remarkable human-likeness (see Katz et al.'s study in this book).

From the viewpoint of human-centred design research the question of robots' human-likeness raises even more issues. Fornari and Congiano's chapter underlines the significance of a robot's face for making human–robot interaction more similar to human–human interaction. While this is definitely true to some extent, another question follows from it: do people like to interact with robots as they interact with people? There are already some studies suggesting instances when people would not like a robot to communicate like people do, but in a more computer-like manner (e.g. Baron, forthcoming). This leads us to propose that more research is needed to understand when to use a human face and when to use a screen as an interface between robot and human user. In this regard, one of the key concepts, as identified in this volume, is the intuitivity of interfaces. It would be incautious to say that a human face is always more intuitive than a screen as we are used to communicating with other humans face-to-face. However, we should not overlook that we are used to communicating with technologies, not through face, but through screens and keyboards (starting from television and home PCs, smartphones and tablet computers), which in the present day are highly integrated and highly intuitive information communication technologies (Emerson 2014). Like Moniz's argues in this book, one of the aims in robot development today is to make user interfaces highly intuitive. If we utilise previous technologies which are already widely adopted and are based on intuitive interfaces between a human user and machine, such as the seminal iPad and other tablet devices that followed, it might smooth the way towards the acceptance of *domestic* social robots in our home and other private spheres.

We finish by considering the implications for further research and the limitations of the research topic explored in this volume examining a human-centred approach to social robots. We have outlined herein the foundations of social robot studies and provided an interdisciplinary collection of cutting-edge articles in the social sciences that combine theory with statistical, experimental studies and data about human users collected from observation as well as participative surveys. The inevitable limitation of this collection of studies is that being studies about social robots at an exploratory and early stage of the research field, the overall discourse can appear a little fragmented. Furthermore, the book has the limitation in that it does not present results from Japan or South Korea which, by many measures, are leading robot societies in the world. Comparative data from these countries would have certainly enriched the contents of the book, yet we decided to give more emphasis on the production of new knowledge considering countries and cultures (Europe, China and the US), which have not received as much scholarly attention as Japan or South Korea, but which are likely to witness an intensive wave of robotification in the future.

Future research needs to explore more about people's imagination, social representations and public awareness of robots in order to grasp better the conceptualization and the attitudes of people towards robots. At the same time, future research needs to address more the behaviours towards and the practices of use of current social robots by different social groups of varying age ranges and life stages such as children, elderly, adults, house persons and service personnel. Additionally within this framework, the study of social robot users' online communities can offer many useful insights towards the design of really user-friendly future social robots.

References

Baron N (forthcoming) Shall we talk? Conversing with humans and robots. The Information Society

Dautenhahn K (2004) Robots we like to live with? A developmental perspective on a personalized life-long robot companion. In: Proceedings of the 2004 IEEE international workshop on robot and human interactive communication, Kurashiki, Okayama, Japan, 20–22 Sept. 2004

Emerson L (2014) Reading writing interfaces: from the digital to the book bound. Minnesota University Press, Minneapolis

Eurobarometer (2012) Public attitudes towards robots. Special Eurobarometer 382: European Commission. http://ec.europa.eu/public_opinion/archives/ebs/ebs_382_en.pdf. Accessed 3 Nov 2014

MIX
Papier aus verantwortungsvollen Quellen
Paper from responsible sources
FSC® C105338

If you have any concerns about our products,
you can contact us on
ProductSafety@springernature.com

In case Publisher is established outside the EU,
the EU authorized representative is:
Springer Nature Customer Service Center GmbH
Europaplatz 3, 69115 Heidelberg, Germany

Printed by Libri Plureos GmbH
in Hamburg, Germany